The Institute of Biology's
Studies in Biology no. 16

Plant Symbiosis

by G. D. Scott B.Sc., Ph.D.
*Senior Lecturer in Plant Physiology,
University College of Rhodesia, Salisbury, Rhodesia*

New York · St. Martin's Press

© George D. Scott, 1969

First published 1969

*First published in
the United States of America in 1969*

*First published in Great Britain by
Edward Arnold (Publishers) Ltd.*

Library of Congress Catalog Card Number: 70-83204

Printed in Great Britain by
William Clowes and Sons Ltd, London and Beccles

General Preface to the Series

It is no longer possible for one textbook to cover the whole field of Biology and to remain sufficiently up to date. At the same time students at school, and indeed those in their first year at universities, must be contemporary in their biological outlook and know where the most important developments are taking place.

The Biological Education Committee, set up jointly by the Royal Society and the Institute of Biology, is sponsoring, therefore, the production of a series of booklets dealing with limited biological topics in which recent progress has been most rapid and important.

A feature of the series is that the booklets indicate as clearly as possible the methods that have been employed in elucidating the problems with which they deal. There are suggestions for practical work for the student which should form a sound scientific basis for his understanding.

1969

INSTITUTE OF BIOLOGY
41 Queen's Gate
London, S.W.7.

Preface

Coexistence of organisms is so nearly universal that no student can claim to be fully versed in biology until he has at least been introduced to the essential principles of symbiosis. The student himself is a partner in a complex cyclic symbiosis—as, indeed, are most members of the animal kingdom. Unhappily, space does not permit a parallel treatment of these aspects of symbiosis. But the implications of plant symbiosis are nonetheless important. In the long term, the author can foresee the creation of new symbiotic systems as one means towards increasing the productivity of land and water.

In the 'symbiosis' that has arisen in recent years between the separate disciplines of botany and zoology in schools and particularly in universities, it is to be hoped that one of the 'unique processes' (see Chapter 4.2) will be the development of courses specifically designed to provide a theoretical and practical knowledge of symbiosis. If this book encourages even a few institutions to adopt such courses, its purpose will be adequately served.

Salisbury, Rhodesia, 1969   G.D.S.

Contents

1	The Concept of Symbiosis	1
2	The Origin and Development of Symbiosis	4
3	Morphological Integration of Symbiotic Systems	7
	3.1 The lichen symbiosis	8
	3.2 Nodule systems	12
	3.3 Mycorrhizal systems	17
	3.4 Endophytic and endozoic algal systems	22
	3.5 Integration of multiple symbionts	24
4	Physiological Integration	28
	4.1 Interflow of metabolites	28
	4.2 Unique physiological processes	39
	4.3 Adaptive physiological processes	44
	4.4 The integrated symbiotic unit	52
5	Symbiotic Systems in Nature	53
	5.1 Symbiosis in the economy of Nature	53
	5.2 The struggle for existence	54
References		57

The Concept of Symbiosis   1

It is an axiom of life that the individuals of a plant species tend to occur in association with each other, in the sense that they show aggregated growth. This is common to all plant species and is caused partly by limitations on diaspore dispersal, partly by limitations on competitive ability. These factors, operating on plants wherever they grow, are responsible for the spatial aggregation of plants into communities. They grow in communities because it is in such communities that their chance of survival is greatest. Plants are 'forced' into association by their environment.

This is the concept on which we must base our consideration of symbiosis —the fact that plants, like or unlike, tend to grow as groups of individuals rather than in random distribution. Grouping is ultimately a physiological phenomenon; all habitat preferences can be related to the satisfaction of physiological requirements.

Within the compass of universal association there are various degrees of interdependence between species. Many plant communities contain species that are regarded as exclusives; they occur only in these communities and never in any others. This is the most easily recognizable base line of physiological interdependence. It is the logical starting point of a series of plant forms showing increasing specialization of habitat and of physiological function.

There are various conditions of coexistence within this series, such as epiphytism involving extreme specialization of habitat, and partial parasitism in which one of the participants is deficient in certain physiological attributes but whose requirements are met by its co-participant. Genera such as *Bartsia*, *Euphrasia* and *Striga* contain numerous species of this character. Further loss of physiological attributes is seen in the so-called saprophytic orchids and in genera such as *Cuscuta*, *Orobanche* and *Monotropa* that are devoid of chlorophyll, or nearly so, and are dependent upon chlorophyllous plants for their carbon nutrition.

The series culminates in absolute mutual interdependence. Physiological specialization has advanced so far that the species involved have become physiologically incapable of free existence except in laboratory culture: but in association as a symbiotic system, each contributes to the other the physiological requirements that are deficient.

Somewhere within this hierarchy of interdependence we must be able to assemble criteria that will allow us to arrive at a satisfactory formulation of symbiosis. The original meaning of symbiosis, conveyed by DE BARY in 1879, was simply *the living together of dissimilar organisms*. His definition was framed on his comprehensive knowledge of association between plants and between animals, and particularly on his own researches on lichens, for it was he who first commented on their dual nature, five years before the classic communication of SCHWENDENER (1869).

Where then, do we draw the line of separation between symbiotic and non-symbiotic association, bearing in mind the virtual certainty that no living organism can exist under natural conditions without being in some way physiologically associated with one or more other organisms?

DE BARY's broad concept of symbiosis clearly did not include associations in which interdependence is only of a secondary nature; nor did he consider chance associations of an ephemeral nature, such as may occur at any point in the lifetime of an individual plant, to be symbiotic. One of the basic criteria of symbiosis should thus be that the association is a **permanent feature of the life cycles of the organisms** and not a casual association that may or may not take place according to environmental circumstances. On this criterion we are able to exclude, for the present at least, the numerous association complexes in plant and soil communities, including epiphytism.

Permanent association implies **physical contact between the participating organisms**. Symbiosis and 'free' existence are thus conveniently separated into *mutual contact* associations, falling within the scope of symbiosis, and non-contact or *distant* associations.

Physical contact between organisms implies physiological interplay either in the form of **unilateral or bilateral movement of metabolites**, or in the form of **amelioration of environmental status**. The latter occurs when an organism, by permanent association with another, increases its ecological amplitude to that of its partner, or when the holobiont (the complete symbiotic unit) shows an extension of amplitude compared to that of either of the symbionts in free-state existence.

Physiological interplay, in addition to sustaining either or both organisms, greatly increases the chance of **morphogenetic effects** which may have a physiological basis; it also provides the opportunity for the **production of metabolites that are not formed by either of the organisms separately**.

An association between organisms that is characterized by at least four of these six attributes may thus be considered to constitute a symbiosis. But within this definition, parasitism finds a place by virtue of unilateral transference of metabolites and increase in ecological amplitude. There can be no doubting the physiological advantage acquired by evolution of the parasitic habit (here, I refer to crude, unspecialized parasitism). This is a 'once for all' advantage, however, and it cannot be furthered without loss of parasitic status. How far parasitism evolves towards the condition of mutual tolerance is measured by the resultant of the factors of compatibility and of incompatibility. Benign parasitism is the result of a condition of stalemate between host and parasite. There are many such conditions of relatively innocuous parasitism, but the inescapable trend is towards elimination of the host and therefore towards self-destruction.

Consider, for a moment, a fungal parasite of a higher plant. The most vigorously parasitic strain of that fungus has an evolutionary advantage and therefore selection occurs in its favour. On the other side of the fence, so to speak, the most vigorously resistant host strain likewise has an evolutionary advantage, particularly if the parasitized state results in diminution of fertility. The interplay of these opposing trends dictates the success of the host over the parasite or *vice versa*. But whichever way we look at it, the parasite cannot win, though it may do so in the short term. Selection in favour of the parasite leads to elimination of the host and therefore to self-elimination: selection in favour of resistance to the parasite leads to elimination of the parasite.

This, then, is the fundamental difference between parasitism and other conditions of association. Parasitism is unique in the world of plant life—DE BARY described it as the most exquisite example of symbiosis—but when we note that the relationship between host and parasite is essentially one of unilateral or non-equilibrated symbiosis, there is room for doubt about its inclusion within the restricted concept of the term.

Symbiosis, consequently, is perhaps best defined as **a state of equilibrated physiological interdependence of two or more organisms involving no permanent stimulation of defensive reaction mechanisms.**

While many of the systems of coexistence between plants, and between plants and animals, are covered by this concept of symbiosis, there are many others of a transitional nature which, with our present knowledge of their physiology, defy classification either as casual, parasitic or symbiotic relationships. It is these intermediate systems—the lichen-like associations of algae and fungi, the many casual associations between fungi, bacteria and seaweeds, and between micro-organisms in the soil and higher plants—that reveal the existence of a continuum, and a perpetual state of flux, in the physiological interrelationships between organisms.

No longer can we think in terms of symbiosis being the sacrosanct appellation of a few highly integrated systems of coexistence, such as lichens, legume root nodules or hydroids. These are but the end-points of several separate lines of evolution that had their beginnings in casual 'acquaintance', such as we can discern in so many other spheres at the present time. The finer limits of encompassment of the term symbiosis are really unimportant and meaningless—it is sufficient that we think of it as a physiologically induced relationship, and keep in mind the concept of the 'ideal' symbiosis.

The Origin and Development of Symbiosis 2

Present-day symbiotic systems have obviously been derived from free-living organisms and it is assumed that they have developed from casual relationships between the organisms concerned. But there is no means of telling how the casual associations took place or how they progressed along the path of increasing complexity. We can only surmise that, far back in evolutionary time, certain pairs of organisms were favoured in the process of natural selection by virtue of the fact that one or both could obtain required nutrition with the expenditure of less energy than in free-state existence. Alternatively, mutational loss of a cellulolytic (cellulose hydrolysing, or cellulose splitting) enzyme system may have induced one organism to obtain its carbon nutrition in the form of a simple carbohydrate from another organism: for plant metabolism is bound by the tenet that if a soluble carbohydrate is available to a carbon heterotrophe, it will be metabolized 'in preference' to an insoluble higher form that requires an extracellular enzyme system to break it down to a soluble, translocatable form.

Since the majority of symbiotic systems include at least one carbon heterotrophe, it follows that the major lines of evolution have been towards the sharing of radiant energy fixed by one of the participants. But carbohydrate nutrition is only one of the possible starting points of symbiosis. Many micro-organisms are known to be incapable of synthesizing certain other metabolites that are required for their growth processes. One of the most common deficiencies in this respect is failure to synthesize either pyrimidine or thiazole, the two precursors of thiamin (vitamin B_1). Two familiar examples are *Mucor ramannianus* and the wild yeast *Rhodotorula rubra*. Neither of these will grow in pure culture without an external source of thiamin or its precursors because *M. ramannianus* cannot synthesize thiazole and *R. rubra* cannot synthesize pyrimidine. Since each possesses half the ability to synthesize thiamin, we have in this pair of organisms the potentiality for its complete synthesis. This in fact does take place when they are grown in mixed culture. Synthesis of thiamin by each organism is made possible by uptake, from its neighbour, of the moiety that it cannot synthesize.

Numerous other pairs of organisms can grow successfully in mixed culture, for example the fungus, *Polyporus adustus*, and the yeast *Nematospora gossypii*. The former is deficient for thiamin and the latter for biotin. When grown together, mutual interchange of the two vitamins or their precursors renders the system self-sufficient for both.

A more complex system is the cyclic interchange of metabolites between three organisms. This has been demonstrated (DAVIS, 1950) for mutant strains of *Escherichia coli*, an intestinal bacterium, which showed blockages at successive steps in the synthesis of the amino acid arginine.

Precursor → Ornithine → Citrulline → Arginine

Davis's mutant 1 (Fig. 2-1) required an external source of ornithine for completion of the synthesis; mutant 2 required citrulline and mutant 3 arginine itself. When the three mutants were inoculated side by side on the same culture plate, the growth of mutants 1 and 2 was stimulated by excretion of citrulline by mutant 3. This excretion is apparently the result of inability to convert citrulline to arginine. Ornithine, accumulated by mutant 2, was likewise excreted and stimulated growth of mutant 1.

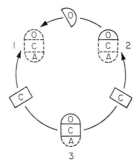

Fig. 2-1 Schematic representation of the probable pathway of interchange of ornithine (O) and citrulline (C) between mutant strains of *Escherichia coli* leading to synthesis of arginine (A). Dotted lines represent inability to synthesize the amino acid concerned (DAVIS, 1950, *Experientia*, **6**, 41–50).

The most pertinent feature of such 'artificial' symbiotic systems is that growth becomes possible as a unit, whereas in separate culture, growth is inhibited by failure to synthesize required metabolites. This is the exact equivalent of supplying the deficient metabolites in the growth medium, and is totally independent of physical contact between the organisms. Provided that associated growth is not inhibited by excretion of antibiotics or by other defence reactions, it is possible to create a 'symbiosis' between any two micro-organisms, each deficient for an intermediary metabolite that is synthesized by the other.

By following the techniques outlined in No. 2 of this series (JACKSON and RAW, 1966), a variety of soil fungi and bacteria can be isolated and tested for vitamin or amino acid requirements. Various combinations of these micro-organisms may then be plated out together as a simple but highly instructive exercise in the creation of 'artificial' symbiosis of the linear or cyclic form.

Such elementary forms of symbiosis involve no tropic, tactile or chemical stimuli or responses. They are no more than the fortuitous result of two or more organisms being brought within exudate diffusion range of each other —a situation that is highly probable in the majority of soils. LOCHHEAD and BURTON (1957) have shown, for example, that as much as 27% of the bacterial flora of a soil under barley is deficient for one or more growth substances. On this basis, it is virtually certain that a 'pool' of growth substances or their precursors serves to maintain growth of these microorganisms in their natural habitats.

A mutant strain of an organism involving loss of physiological attributes is faced either with extinction or with the possibility of survival by means of association with another organism from which it can make up its physiological deficiencies. The kingpin on which successful association rests is the inherent propensity for organisms to exude a variety of substances from their cells. Chance growth of two organisms within mutual exudate diffusion range provides the first stepping-stone towards symbiosis. What was initially a casual association becomes selective; this in turn becomes obligatory following the loss of autotrophism in respect to various intermediary metabolites.

From such beginnings, illustrated by the numerous present-day 'distant' associations prevalent in the exosphere* of most plants, symbiotic systems are presumed to have developed to their full complexity of physiological and morphological relationships.

It is but a small step from the 'distant' casual association to the permanent 'contact' relationship, in the evolutionary scale of time. In physiological terms this is a major step beset with barriers. The greatest barrier is that of antibiosis—the antithesis of symbiosis. But it is also the principal factor which ensures that, of all the species and physiological strains of potential symbionts (symbiotic organisms) growing within mutual exudate diffusion range, only the physiologically 'right' ones will succeed in establishing a permanent relationship. There are many examples, in all fields of plant association, of the antagonistic and occasionally lethal effects of physiological interaction. Undoubtedly there are many 'tries' at symbiotic association, but it is only the select few with the requisite physiological compatibility that can attain this status.

*The exosphere of a plant is that volume of space surrounding it that roughly delimits the exudate diffusion range. The term is applicable to all living organisms; the rhizosphere, rhizoplane and phyllosphere are constituent parts of the exosphere.

Morphological Integration of Symbiotic Systems 3

Practically all symbiotic systems are characterized by the fact that one of the symbionts is the morphologically dominant partner of the association. That is to say, in terms of size or volume, we can recognize a macro- and a micro-symbiont. This has no bearing on the physiological activities of the co-symbionts; in fact the morphologically major partner is in many cases the physiologically minor one.

In morphological terms, one of the symbionts is either entirely enclosed within the tissues of the other or is partly exposed to the substrate or environment. A few examples of apparently total intracellular enclosure are to be found in the endozoic symbioses and a very few also in what are termed syncyanoses, or endocellular associations, involving blue-green algae. All other systems are characterized by extracellular contact in which there is either total or partial enclosure.

The distinction between intra- and extracellular enclosure is far from being precise. It is only recently, for example, that *Rhizobium* bacteroids were shown to be enclosed within an envelope in legume nodule cells, this envelope being continuous with the plasma membrane (BERGERSEN and BRIGGS, 1958). Similarly, in other symbiotic systems including lichens, fungal haustoria are in some cases now known to be enclosed within the co-symbiont cell membrane, not, as was formerly thought, in direct contact with the cytoplasm. It is also of interest that modern electron microscopy has shown that the haustoria of certain fungal parasites do not penetrate the host cytoplasm, but are enclosed within a membrane that may be of host origin.

It remains to be seen whether further cytological investigation of co-symbiont relationships will reveal that cell membranes act as the boundaries between all symbionts and that there is thus no true intracellular association. From the point of view of physiological equilibrium—one of the basic criteria of symbiosis—it is certain that gross derangement would occur if there were no physiological barriers, in the form of membranes, between the co-symbiont protoplasts.

Permanent contact between different organisms implies that, to some extent, an integral unit is formed. That this unit frequently functions as an apparently single morphological entity is illustrated by many systems, for example lichens (§3.1), root nodules (§3.2), *Cyanophora* (a flagellate containing two or more blue-green algal symbionts), hydroids and many other coelenterates (§3.4). The common denominator of systems such as these is that one of the symbionts is completely removed from physical contact with the substrate or environment. It is therefore entirely dependent upon its co-symbiont for nutrient and water supply at least, and in some cases also for carbon supply.

Physical enclosure of one symbiont entails a change in shape or volume of the co-symbiont, but this does not in any way account for the very pronounced changes in morphology, such as the advent of unique morphological structures, that are so characteristic of symbiotic systems. These changes, with the assumption of symbiosis, are the morphological expression of physical and physiological interaction between the symbionts.

3.1 The lichen symbiosis

The association of alga and fungus in the lichen symbiosis has resulted in the evolution of numerous lines of morphological construction that are quite unique to the symbiotic condition. Figure 3–1 shows two major types of morphological form recognizable in lichens. Because of this unique morphology, there has arisen an assemblage of terms descriptive of the lichen thallus and of the relationship between the symbionts.

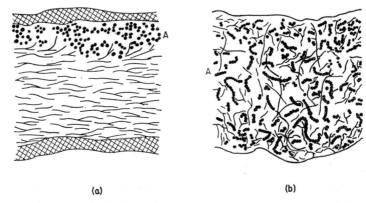

Fig. 3–1 The heteromerous, or layer-differentiated lichen thallus (**a**), compared with the homoiomerous, or relatively undifferentiated thallus (**b**). Physiologically, the important difference between the two types is the distribution of the photosynthetic cells, A.

The primitive or homoiomerous type of thallus construction reveals an undifferentiated mixture of algal cells (or filaments) and fungal hyphae, while the more advanced or heteromerous construction more nearly resembles that of the angiosperm leaf. The phycobiont (algal symbiont) in this case is confined to a specific region of the mycobiont (fungal symbiont) tissue, usually as a sub-cortical layer.

Most lichen mycobionts are ascomycetes and, as such, reproduction is effected by means of ascospores produced in apothecia of varying shape.

No lichen phycobiont is known to reproduce by sexual means in symbiosis, although they may do so in pure culture. This is an obvious weak point in perpetuation of the species because the chances of any ejaculated ascospore coming into contact with a compatible phycobiont cell are in the realms of several millions to one. It so happens, however, that vegetative reproduction of lichens is a much more effective means of dispersal than the initiation of new symbioses from ascospore and phycobiont cell. Thallus

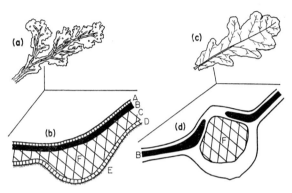

Fig. 3–2 The *Sticta* 'leaf', (a) and (b), in comparison with a typical dicotyledonous leaf, (c) and (d). Upper and lower cortex, A and D; photosynthetic layer, B; medulla, C; rhizinae, E; midrib, F.

fragmentation and dispersal by environmental agencies is perhaps the most effective means. Many species, however, produce structures termed soredia which function partly as dispersal organs and partly as a means of thallus aeration. The soralium, comprising a mass of individual soredia, is a powdery pustule on the lichen thallus, rather akin to the angiosperm lenticel. Soredia are non-wettable and are readily blown or shaken off the thallus.

Morphological integration has reached a high level in the lichen symbiosis: so much so that we can draw a remarkably close parallel between the heteromerous lichen thallus and the angiosperm leaf. There is a most striking resemblance, for example, between the 'leaf' of *Sticta filicina* and the conventionalized dicotyledonous leaf (Fig. 3–2). The *Sticta* leaf is a dorsiventral lamina with a thickened central region or 'midrib' and is supported above the substrate by a stem-like structure. Its photosynthetic tissue—in this case the phycobiont layer—occupies a position equivalent to that of the palisade layer of the angiosperm leaf, immediately underneath the 'epidermis' of several layers of isodiametric cells. Numerous pores (cyphellae) are present on the lower surface of the lamina. Physiologically, they provide for aeration and to this extent they are analogous to stomata; in structure and distribution frequency, however, they more nearly resemble the angiosperm lenticel.

The phycobiont layer in *Sticta*, and indeed in the majority of lichens, has become relegated to a physiologically defined position in the thallus. It is thought to develop in this characteristic position largely in response to light intensity control, although growth regulators such as indole acetic acid, gibberellic acid and phytokinins may also be involved.

When a lichen symbiosis arises *de novo* or from a soredium, it does so as a completely undifferentiated mixture of algal and fungal cells. This undoubtedly takes place in cracks and crevices of the substrate that are protected from direct sunlight. With increase in size, the holobiont becomes raised above the substrate level and is exposed to the morphogenetic influence of light. In the juvenile undifferentiated holobiont, the phycobiont cells close to the surface of the unit are subjected to the bleaching effect of full sunlight. This is a well-known phenomenon in laboratory cultures of green algae and it is usually found that they become light-saturated at intensities well below that of maximum sunlight.

It is now believed (AHMADJIAN, 1967) that many lichen phycobionts cannot adequately withstand the pressures of the environment in the free-living state. Species of the genus *Trebouxia*, for example, which are the phycobionts of the majority of 'green' lichens, are seldom recorded amongst algal covers on soil, rocks or tree bark. In our present state of knowledge of algal symbionts it is unwise to make any generalization regarding their capacity to exist in the free-living condition, but we must note that situations similar to that of *Trebouxia* have been reported for the phycobionts of numerous endozoic systems. Also, in those systems that have reached the state of near-ideal symbiosis, such as *Paulinella chromatophora* (a freshwater rhizopod), there is little physiological basis for supposing that these phycobionts can exist except in the symbiotic condition.

Increasing exposure to sunlight, during development of the lichen thallus, is thus thought to contribute to relegation of the phycobiont layer to such a position that the cells can survive and multiply in phase with the mycobiont along the meristematic margins.

The depth of the phycobiont layer is also thought to be under light intensity control. While the outer limit of the layer is defined by sensitivity to high light intensity, the inner limit can be considered to be equally sensitive to low light intensity. It is defined by that position within the holobiont at which light intensity becomes too low for survival of the individual cells. This limiting effect of light intensity can be demonstrated by culturing lichens, such as *Peltigera*, upside down under laboratory conditions. The phycobiont cells grow upwards through the medulla in response to the increased light intensity at the morphologically lower limit of the phycobiont layer.

HILL and WOOLHOUSE (1966) have shown that there are significant differences in thallus thickness between specimens of *Xanthoria parietina* growing on shaded tree trunks and on exposed maritime rocks. Their data, part of which is reproduced in Table 1, demonstrates that the phycobiont layer is thicker in shade thalli than in sun thalli. Thus, low light intensity, in accordance with the above theory, appears to induce thickening of the phycobiont layer.

Table 1 Vertical thickness (in microns) of the tissue layers of *Xanthoria* thalli on trees and on maritime rocks. (From Table 3 of HILL and WOOLHOUSE (1966); courtesy of *The Lichenologist*.)

Habitat	Upper cortex	Algal layer	Medulla	Lower cortex
Trees	17·6 ± 1·3	46·4 ± 6.0	52·1 ± 3·2	21·5 ± 1·7
Rocks	25·4 ± 1·2	34·2 ± 3·2	111·9 ± 7.1	27·3 ± 1·3

This effect cannot be divorced from the light-screening effect of pigmentation of the upper cortex. Shade thalli have a much lower level of pigmentation than sun thalli. The difference in light intensity reaching the phycobiont layers of the two types of thalli is thus not so great as the actual difference in light intensity at the surface of the thalli, but is sufficient to promote the observed morphogenetic effects.

This is not the complete story, however, for it is known from field observations and from the work of HILL and WOOLHOUSE (Table 1), that the lower cortex, the medulla and the upper cortex of lichens increase in thickness when exposed to full sunlight and are thinner than normal when grown in low light intensity. This light-induced morphogenetic effect on the mycobiont is as inexplicable as the corresponding effect seen in sun and shade leaves of angiosperms.

Stratification of the lichen thallus is no isolated feature of a few foliose forms. It is characteristic of all except the gelatinous lichens and a few primitive forms. Even in *Collema* and *Leptogium*, the two major gelatinous genera, there is a series ranging from the homoiomerous to the near-heteromerous condition. Relegation of the phycobiont layer to the sub-cortical position is not in any way related to evolutionary lines within the lichens—cogent argument for the participation of light intensity (a factor common to all) as a major morphogenetic factor.

The consequence of this simulation of the angiosperm leaf is that the lichen thallus has become an efficient photosynthetic structure. This is perhaps one of the clearest examples of homoplastic development amongst plants; the remarkable feature in this case being that it is achieved by *two* organisms functioning as an integral morphological unit.

Mention may be made here of the recent demonstration (SCOTT and SMILLIE, 1967) that the DNA of *Euglena* chloroplasts codes for RNA of the chloroplast ribosomes. This DNA-RNA system and its dependent protein synthesis appears to be independent of nuclear DNA. The ability of chloroplasts to synthesize their own protein systems by a coding sequence distinct from that for cytoplasmic proteins, implies that they have at least partial genetic autonomy. This lends weight to the recently revived theory that chloroplasts have perhaps evolved from some form of symbiotic bacterium. If so, then the most successful groups of plants—those that have attained autotrophy by virtue of their ability to convert radiant into chemical energy—will be seen to have achieved this through symbiosis.

If the chloroplasts of green algae are thought of as having a similar origin, the trends seen in present-day green lichens, and in other symbiotic systems containing green algae, pose the interesting problem of a phycobiont, formerly 'free living' but whose chloroplast is derived from a symbiotic bacterium, embarking on a second experience of symbiosis. In the heteromerous lichen thallus, evolution would be seen to have resulted in the construction of a model of the angiosperm leaf using an already symbiotic system as the 'chloroplasts'.

Blue-green algae are commonly regarded as having strong affinities with bacteria, and indeed have been said to stem from them. Blue-green lichens may thus be regarded as a further pathway by which the ultimate in photosynthetic efficiency is in process of being achieved.

The case of *Cyanophora* (Plate 5) appears even more complex in this light, for the blue-green phycobiont cells have been shown to have lost all semblance of a cell wall (HALL and CLAUS, 1963). *Cyanophora* is a symbiotic system that is perhaps advancing to the 'asymbiotic' state of perfect integration by evolutionary loss of phycobiont autonomy.

If we project the path of evolution along these lines, perhaps WALLROTH's (1825) idea of the phycobiont cells of lichens being an integral part of the mycobiont (although thought by him to be of reproductive function) may not be so erroneous as we consider it to be at the present time.

3.2 Nodule systems

Many Angiosperm and a few Gymnosperm species, in association with micro-organisms, produce nitrogen-fixing root, leaf or even stem nodules. Root nodules are a characteristic feature of the Leguminosae, but are also found in a few other families such as the Rosaceae, Ericaceae, Betulaceae and Elaeagnaceae and in the gymnospermous families Podocarpaceae and Taxaceae. The majority of species in the sub-families Papilionoidae and Mimosoidae of the Leguminosae are known to nodulate; the most common examples are peas, beans, clovers and lucerne. Legume root nodules have been known for more than eighty years to fix molecular nitrogen. This is the only known instance of symbiotic association of a bacterium—in this case *Rhizobium*—with a higher plant, that has evolved the unique symbiotic ability to fix nitrogen. Neither symbiont in the free-living state is known to have this capacity.

Nitrogen fixation by leguminous plants can be conveniently demonstrated by growing plants from seed in nitrogen-free water- or sand-culture

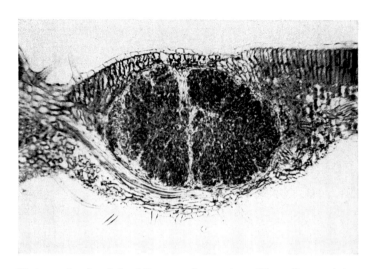

Plate 1 Leaf nodule of *Pavetta schumanniana*. The cells containing the bacterium are surrounded by an epithelium. (×65)

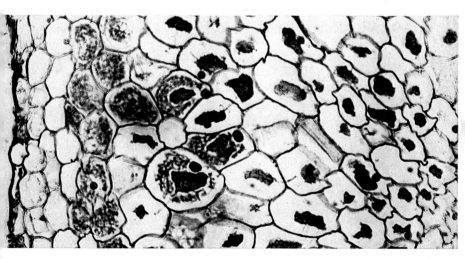

Plate 2 Rhizome of *Corallorhiza* showing the 'host' cells in the outer cortex and the 'digestive' cells in the inner cortex. (×150)

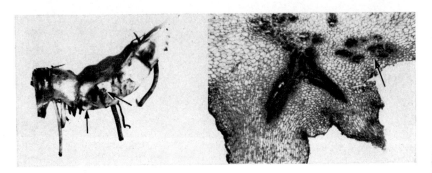

Plate 3 Rhizome of *Gunnera perpensa* showing the groups of *Nostoc* in relation to the adventitious roots (arrowed). In longitudinal section, the *Nostoc* colonies are seen to occur in the middle cortex. (Rhizome, × ½; L.S., × 30)

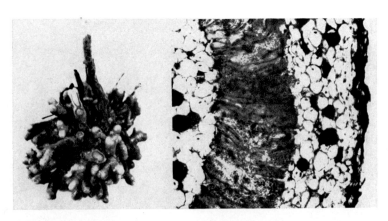

Plate 4 The symbiotic coralloid roots of *Cycas revoluta*. In transverse section the *Nostoc* colonies are seen to occupy a zone surrounding the stele. (Coralloid roots, × 5/4; T.S., × 60)

and inoculating the culture medium of one set of plants with crushed nodules from the same species. The marked difference in growth between the control plants and nodulated plants provides striking indirect evidence that the nodules supply sufficient nitrogen to maintain normal growth.

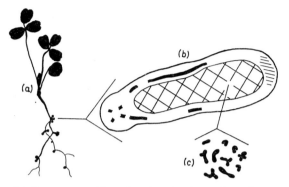

Fig. 3-3 (a) Lucerne seedling showing the formation of root nodules. (b) Section of a lucerne nodule and root. The bacteroids, (c), are confined to the intrastelar tissue.

Direct evidence of nitrogen fixation is more difficult to achieve. Nitrogen gain can be estimated by the Kjeldahl method of nitrogen determination, but this is limited in accuracy and cannot be used for short-term experiments. An alternative method is to expose the root systems of nodulated plants for a few hours to an enclosed atmosphere containing the isotope $^{15}N_2$, followed by mass spectrometric evaluation of the ratio $^{15}N:^{14}N$ in the plant tissue. A recent, but indirect, method relies on the fact that the nitrogen-fixing system has a higher affinity for acetylene than for nitrogen. Acetylene is preferentially reduced to ethylene which can be estimated with a high degree of accuracy by gas chromatography (STEWART et al., 1967).

The structural detail of a typical legume nodule is shown in Fig. 3-3. The nodule is initiated by the entry of *Rhizobium* cells from the soil through a root hair or through broken cells of the piliferous layer. An 'infection thread' develops which is initially enclosed by an invagination of the macro-symbiont cell wall. It eventually penetrates the apical part of the invagination and becomes surrounded by a membrane system of macro-symbiont origin. The 'infection thread' continues inwards through the cortex until it reaches a cell, larger than its neighbours and having double the somatic number of chromosomes. Such tetraploid or octaploid cells, as the case may be, are the presumed starting points of root nodule formation. With its immediate neighbours, this cell is stimulated to repeated division by a change in growth regulator equilibrium. Indole acetic acid (IAA) and phytokinins are probably involved in this process. The former at least is known to be formed by *Rhizobium* from tryptophane of macro-symbiont origin.

Initial development of the nodule is closely similar to that of a lateral root, with the exception that from an early stage, the central cells become filled with 'packets' of *Rhizobium* cells enclosed within a membrane system. Electron micrographs show that these membranes are almost certain to be invaginations of the macro-symbiont cytoplasmic membrane; the *Rhizobium* thus remains extracellular (Chapter 3).

In the early development of the nodule, the *Rhizobium* cells become enlarged and changed in shape to form 'bacteroids' or bacteria-like cells. They are typically 'X' or 'Y' or boomerang-shaped (Fig. 3–3). It is not known how this bacterial morphogenesis takes place. The process involves both symbionts, for during the conversion phase from rod to bacteroid form, the enveloping macro-symbiont cells undergo considerable hypertrophy. The change may be partly caused by an increase in bacterial membrane permeability induced by a macro-symbiont metabolite. The resultant shapes of the bacteroids may then be due to mutual pressure during increase in volume.

Apart from the hypertrophied cells containing the bacteroids, nodule construction follows the general plan of a lateral root, with the exception that there is no root cap and the apical meristem is of more limited life. The nodule, arising from the morphological and physiological integration of the two symbionts, is thus not a unique entity of symbiosis; it is based on a structure—the lateral root—whose formation is within the normal genetic capability of the macro-symbiont. Its limited axial growth and its radial thickening are the outcome of some disarrangement in growth regulator equilibrium induced by the presence of the micro-symbiont. Although its embryogeny is not that of a lateral root, subsequent development of the nodule parallels that of a lateral root so closely that it cannot be regarded as a new morphological entity arising from symbiosis.

Such is the general pattern of development of the legume nodule, but it varies in several respects between species. Branching of the meristematic apex is a commonplace feature of numerous nodules and, in some species, leads to the formation of fairly large nodule clusters. All legume nodules, even those that are perennial, are of limited life. Their senescence phase culminates in the breakdown of the nodule tissues, lysis of the bacteroids and liberation of previously dormant rod forms of the micro-symbiont to the soil. These become part of the pool of *Rhizobium* cells in the soil that initiates new nodules on legume plants. (See STEWART, 1966, for further details of legume nodule structure.)

The legume-*Rhizobium* association is not of the permanent nature of the lichen symbiosis and others. Every new root hair formed by the macro-symbiont provides the opportunity for fresh initiation and each new nodule formed may or may not function throughout the lifetime of the macro-symbiont. For any individual legume plant, there can thus be a continuum of new symbioses developing with continued growth of the root system, subject to any inhibitory effect that already existent nodules may have on the formation of new ones on the same root system (NUTMAN, 1952).

Normally, this is not an obligatory symbiosis, but under conditions of extreme soil nitrogen deficiency neither symbiont has much chance of surviving, while in symbiotic association normal growth takes place. The association between legume and *Rhizobium* not only overcomes carbon heterotrophism on the part of the micro-symbiont, but renders the holobiont entirely independent of combined nitrogen supply from the soil.

This is also true of the non-legume plants that bear root nodules. These include species of *Alnus, Coriaria, Hippophaë, Myrica* and several other Angiosperms. A few Gymnosperms, such as *Agathis, Dacrydium, Microcachrys, Podocarpus* and *Saxegothaea*, are also known to nodulate but their nitrogen-fixing ability has not been fully investigated. Two recent additions to the list of nitrogen-fixing non-legumes are *Dryas drummondii* and *Purshia tridentata*, both in the Rosaceae (LAWRENCE et al., 1967; WEBSTER et al., 1967).

The identity of the micro-symbionts of non-legume nodules is still uncertain. There is strong evidence that in *Alnus, Hippophaë* and *Myrica* it is an actinomycete, while in *Podocarpus* it is thought to be a phycomycete. *Agathis australis* has been shown to associate with two micro-symbionts, one thought to be a phycomycete, the other an actinomycete.

Apart from the different micro-symbiont, the non-legume nodule is morphologically quite similar to the legume type. It arises as a lateral outgrowth from a root, subsequent to initiation of the symbiosis. Because nodulated non-legumes are woody perennials, the nodules are usually found in clusters, often very large as in *Alnus*, that persist for several years. Clustering is caused by repeated branching of the nodules to give rise to a spherical mass of nodular tissue of which only the outermost parts are active in fixation of nitrogen. Mature nodules, in keeping with their perennial nature, are ensheathed by a cork layer.

Little is known of the initiation of non-legume nodules apart from suggestions that the process may be similar to that of the legume nodule. 'Infection threads' of the micro-symbiont have been reported, but it is not known whether lateral outgrowth of the embryonic nodule is initiated in a tetraploid cell or whether from a normal lateral root meristem.

The morphological relationship between the symbionts differs in several respects from the legume-*Rhizobium* association. The micro-symbiont is normally present within the cells of the nodule cortex only, enclosed within a membrane system possibly of macro-symbiont origin. It frequently loses its filamentous character and assumes a form similar to the bacteroid form of *Rhizobium* in the legume nodule.

There is a definite zonation of tissues within the nodule, more evident in the longitudinal than in the transverse plane. Cells near the nodule apex contain obviously active hyphae and characteristic hyphal vesicles. Nearer the base of the nodule and towards the centre, in the transverse plane, there is an increase in the proportion of digested hyphae and of bacteroid cells. In the mature basal region, the hyphae have been fully digested or liberated to the soil by cell rupture. The formerly active macro-symbiont cells now appear empty. With continued growth of each nodule, the micro-symbiont penetrates newly formed cortical cells; this process continues for the lifetime of the nodule cluster. (See BOND, 1963, for further details of non-legume nodule structure.)

Digestion of micro-symbiont hyphae is a characteristic of the endotrophic (occurring within the plant) mycorrhizal symbiosis. The occurrence of this phenomenon in the non-legume nodule and the confinement of the micro-symbiont to the extrastelar tissues, place the non-legume symbiosis in a position intermediate between the legume and the endotrophic mycorrhizal symbioses (§3.3).

It is to be expected that other intermediates between the extremes will occur, and this has in fact been shown to be so. *Agathis australis*, the kauri pine, is now known to participate in a polysymbiosis. The root nodules of this species have been known for some time, but it has only recently been demonstrated (MORRISON and ENGLISH, 1967) that two micro-symbionts are involved. One of these is of the typical vesicular-arbuscular type, probably an *Endogone* (§3.3); the other is said to resemble an actinomycete. The small amounts of nitrogen fixation that have been detected in *Agathis* nodules have been ascribed to the actinomycete.

Agathis raises the interesting problem regarding which of the two micro-symbionts is responsible for the morphogenesis of the nodule. Is it the supposed actinomycete, which is presumed to have this effect in other non-legume species, or is it the endotrophic mycorrhizal fungus? The existence of non-mycorrhizal nodules, still retaining the ability to fix molecular nitrogen, indicates that the *Endogone* is not the primary morphogenetic agent, although it may have some effect. More detailed investigation of the nodules of *Agathis* and of related species may yet reveal the presence of a bacterium or other micro-organism hitherto undetected, but in the light of present knowledge, there is no answer to this question.

The *Podocarpus* type of nodule has been shown to be weakly active in nitrogen fixation. The micro-symbiont is a non-septate multinucleate hyphal system; this is consistent with its being a phycomycete, but it has not yet been isolated in pure culture so that its true identity remains unknown. If it is a phycomycete and if it is responsible for the nitrogen-fixing property of the nodules, then this is the first instance of a fungus participating in the fixation of nitrogen.

There is a close parallel between the nodules of *Podocarpus* and those of *Agathis*. The micro-symbiont of both is of the vesicular-arbuscular endotrophic type (§3.3), but *Agathis* has also been shown to associate with an actinomycete and to fix nitrogen. There is clearly a need for further investigation of *Podocarpus* nodules and those of several genera in the

Taxaceae, such as *Dacrydium* and *Saxegothaea*, to determine whether they too contain an actinomycete.

Bacterial leaf nodules present some interesting analogies with other types of symbiosis. They are prominent in one family of flowering plants only, the Rubiaceae—*Pavetta* and *Psychotria* being the best-known examples—but they also occur in the genus *Ardisia*, in the Myrsinaceae. This line of symbiotic association is as close to being obligatory as any other known association. In both families in which leaf nodulation occurs, the macro-symbiont is reduced to dwarfism if grown in the asymbiotic condition.

The nodules are confined to the leaf margins in *Ardisia*, but are scattered throughout the mesophyll in the Rubiaceous species. In section (Plate 1), *Pavetta* nodules are seen to occupy the greater part of the depth of the leaf. They frequently develop under a stoma. Although the bacterium is thought to be transmitted via the seed, entry to the leaf while still rolled in the bud is presumably through stomata. The *Psychotria* micro-symbiont, identified by SILVER *et al.* (1963) as a species of *Klebsiella*, is a nitrogen-fixing bacterium. This property, unlike the legume-*Rhizobium* association, is not confined to the symbiotic condition; it is thus not a unique physiological attribute of the symbiosis.

Dwarfism of the macro-symbiont has been said to be caused by lack of growth regulator formation. In symbiotic plants, the micro-symbiont provides the necessary growth regulators for 'normal' growth. Application of IAA to dwarf plants has not, however, produced the expected return to normality, but gibberellic acid is reported to effect some improvement in growth.

The strange feature of this symbiotic condition is that morphological 'normality' is achieved by symbiosis, while gross derangement of morphogenetic processes takes place in the free-living condition. But this can equally be said of the lichen symbiosis—it all depends on what we, the observers, consider to be 'normal'.

3.3 Mycorrhizal systems

Mycorrhiza, meaning 'fungus-root', is the term applied to the very prevalent association of soil fungi with plants of all kinds that are rooted in the soil or, in the case of certain fern prothalli and liverworts, that grow on the surface of the soil.

The most widely known mycorrhizal associations are those between the roots of trees and basidiomycetous fungi, such as *Amanita*, *Russula*, *Tricholoma*, *Boletus* and *Lactarius*, that habitually grow in forest litter. In these ectotrophic mycorrhizal systems, the fungal mycelium forms a mantle or sheath of hyphal tissue surrounding the tree root tips (Fig. 3-4). It also penetrates between the cortical cells to form the so-called Hartig net. In the healthy mycorrhizal root, the mycobiont hyphae forming the

Hartig net seldom show haustorial penetration of the cytoplasm of the macro-symbiont root cells. Normally, the hyphae of the net do not even penetrate the cell walls of the root, but are confined to the region occupied by the middle lamellae of these cells. There is here obviously very close contact between the cell walls of the co-symbionts.

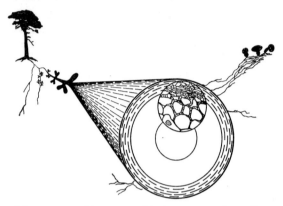

Fig. 3–4 The ectotrophic mycorrhizal system of conifer trees. Short, branched roots of the tree are invested by a mantle of fungal hyphae which ramify sparsely through the forest litter and fructify as the familiar forest-floor agarics. (Shadowgraph of tree adapted from photograph in A. E. HOLDEN, 1952, *Plant Life in the Scottish Highlands*. Oliver and Boyd, Edinburgh. By permission.)

The fungal mantle surrounding the root tips is a felt-like mass of pseudoparenchymatous hyphae extending about a centimetre or more backwards along the root. This fungal investment is so closely knit that it virtually acts as a water sponge. It takes the place of root hairs which are physically, if not physiologically, prevented from growing out.

In comparison with other symbiotic systems, such as lichens and even the various root nodule symbioses, mycorrhizal associations show little morphological integration. Of the many tree species participating in mycorrhizal symbiosis, most seem to be fairly catholic in their 'choice' of associates. That is to say, little specificity seems to be exhibited—apart from a few specialized instances—in the association between tree and fungus.

The most striking difference between mycorrhizal and 'free-living' roots is that the former are shorter, of limited growth, and are more profusely branched. Frequently, the branching is a repeated dichotomy, giving rise to coralloid masses of symbiotic roots. This is characteristic of numerous pine mycorrhizae.

Anatomically, the short roots differ from asymbiotic long roots in several respects. The root cortex cells in close contact with the Hartig net are larger and more radially elongated than their long root counterparts.

Consistent with their limited growth, the mycorrhizal roots have a poorly developed apical meristem, little semblance of a root cap and a scarcely discernible differentiation into longitudinal tissue layers (HARLEY, 1959).

There is no doubt that these differences are the result of a change in the pattern of morphogenesis of the roots and that this is induced by their association with the fungus. Considerable evidence is available (see SLANKIS, 1958) to support the view that one of the major morphogenetic agents is indole acetic acid. Slankis and others have shown that tree roots, isolated in aseptic culture and supplied with exogenous IAA, exhibit many of the morphological characteristics of mycorrhizal roots. The extent of modified morphogenesis is dependent upon the concentration of IAA. At relatively high concentrations (5–10 mg/l), isolated roots of *Pinus sylvestris* are induced to form a high proportion of new roots similar to the short, dichotomously branched, coralloid system of mycorrhizal roots. These roots are radially swollen; they lack root hairs and the cortex cells show a similar degree of hypertrophy to mycorrhizal roots. Slankis has shown that this effect can be terminated by removing the exogenous source of IAA, whereupon the short roots continue growth as normal long roots.

The fact that exogenous application of IAA to roots produces effects similar to those seen in the symbiotic condition, indicates an IAA-producing activity of the mycobiont. Numerous mycorrhizal fungi, including species of *Amanita*, *Coprinus* and *Boletus*, have been shown to produce IAA in pure culture (ULRICH, 1960). We can thus ascribe part, at least, of the morphogenetic activity in mycorrhizal roots to IAA of mycobiont origin.

One significant factor common to all ectotrophic mycorrhizal symbioses is the pattern of morphogenesis in the mycorrhizal roots and in the mycobionts. Taxonomically widely separated mycobiont species induce, in equally widely separated tree species, essentially similar responses—both morphogenetic and physiological. These responses result in the establishment of an easily recognizable morphological entity, yet this entity develops from several hundred different combinations of symbionts. TRAPPE (1962) lists 119 different mycobionts for *Pinus sylvestris* alone. These are also recorded as the associates of about 720 other tree species.

Every 'species' of mycorrhizal association can be compared to the individual lichen species. Physiologically, each is a symbiosis towards the same end—the attainment of autotrophy by association with a photosynthetic organism. The only difference between the two systems, in this context, is the interpolation of the non-chlorophyllous tree tissues between mycobiont and chloroplast in the mycorrhizal symbiosis.

By contrast, there is no such common path of morphogenesis in the vast array of symbiotic systems of the endotrophic mycorrhizal type. Here, with few exceptions, we have a state of symbiosis between a restricted number of fungi, either phycomycetes or fungi imperfecti, and a very wide range of vascular and non-vascular green plants. These include Angiosperms, Gymnosperms, Pteridophytes and Bryophytes.

Mycorrhizal associations of the wholly endotrophic type take place between the fungus *Endogone* (sometimes referred to the form genus *Rhizophagus*) and many different types of green plant. These are non-specialized associations in that there is little morphogenetic response on the part of the macro-symbiont. The mycobiont hyphae ramify throughout the root cortex and produce characteristic vesicles, which are thought to be reproductive and oil storage structures, and arbuscules or 'little bushes' of finely branched hyphae (MOSSE, 1963). It is a characteristic of these vesicular-arbuscular mycorrhizal associations that the mycobiont is eventually digested by the macro-symbiont. This is evidently a type of symbiosis that is not very far removed from antagonistic association, but little is known of its physiology.

There is apparently very little specificity of association between *Endogone* and green plants. It is even recorded as the probable micro-symbiont of the root nodules of gymnosperms such as *Agathis* and *Podocarpus*. But these few instances are so atypical of the better-known relationships between *Endogone* and other plants that it is difficult to accept that it can be the causal factor in morphogenesis of the nodules, even though it be present.

The orchid mycorrhizal systems perhaps provide the best illustrations of close morphological integration at the endotrophic level. For most green orchids, and certainly for all saprophytic orchids, the symbiosis can be considered as obligatory, under natural conditions.

The most common micro-symbiont appears to be *Rhizoctonia*. Numerous species have been isolated from different orchids, and in some instances, two or more strains from a single orchid. There are a few anomalous associations, however, such as that of *Armillaria mellea* and the orchid *Gastrodia elata*. A few other basidiomycetes, including *Fomes* and *Marasmius*, have been reported as orchid symbionts.

Orchids are an excellent example of the success of resorting to symbiosis. They have minute seeds which are virtually incapable of successful germination and further growth unless assisted by an external source of carbohydrates and vitamins. These are conveniently supplied by the mycorrhizal fungus which is present from an early stage in the development of the seedling. The association persists for the lifetime of saprophytic orchids, and for varying periods of time in green orchids, depending on their degree of assimilatory self-sufficiency. In some instances, there is evidence for movement of carbohydrate from the orchid to the fungus, when the former has reached the state of autotrophy.

The many intermediate conditions between saprophytism and autotrophism in orchids portray a range of specialization in the relationship between the symbionts. The complete saprophytes, such as *Corallorhiza* (Plate 2) and *Neottia*, have very much reduced root systems—reduced so much in *Corallorhiza* that roots have never been observed—and an intimate relationship between the majority of the root or rhizome cortical cells and the mycobiont. Concentric zones of the cortex can usually be discerned in which the mycobiont hyphae are either 'active' or in process of digestion. These have been termed the 'host' and 'digestive' layers. In physiological terms, the mycobiont hyphae in the 'host' layer are in stable equilibrium with the cortex cells, but in the 'digestive' layer, they undergo autolysis, or digestion by macro-symbiont enzyme systems.

In those orchids that achieve autotrophy at an early age, the association is of a more ephemeral nature. Root systems are usually well developed and they lack the distinct zonation of the mycobiont in the root cells.

Correlation of morphological integration with the state of autotrophy of the macro-symbiont serves to illustrate, more than in any other group of symbiotic plants, the existence of a mechanism by means of which the tolerance-antagonism system is regulated. The most likely method by which this is achieved is through specific metabolites whose level of production is a function of the photosynthetic efficiency of the macro-symbiont.

Although, for convenience, we divide mycorrhizal systems into endotrophic and ectotrophic types, there is, as in all biological systems, no clear-cut distinction between the two. Numerous intermediates occur and these have been labelled ectendotrophic in recognition of the difficulty of placing them in either category.

One of the best known examples of the intermediate type is the *Monotropa* symbiosis. This plant, commonly known as Yellow Bird's-nest or Pinesap, is confined to forest litter in temperate countries. It is devoid of chlorophyll and relies for its carbon nutrition on a mycorrhizal association that produces a mantle and Hartig net, but which is also endotrophic to the extent that mycobiont hyphae penetrate the epidermal cells.

BJÖRKMAN (1960) has established, by tracer carbon techniques, that the mycobiont of the *Monotropa* symbiosis (thought to be a species of *Boletus*) also associates with neighbouring trees in the formation of an ectotrophic mycorrhiza (see Fig. 4–5). The physiological details of this polysymbiosis will be discussed later (§3.5 and §4.1.2), but it is here noted that a two-way translocation of phosphorus has been demonstrated between the tree and *Monotropa*, in addition to the movement of carbohydrates.

The type of mycorrhiza found in *Monotropa* is typical of many plants in the Ericales, but there is a considerable range from the near-endotrophic to the near-ectotrophic. Some, such as *Calluna*, border on the wholly endotrophic with only a loose weft of hyphae external to the root cortex; others, such as *Arbutus*, lie at the opposite end of the range with a well-defined mantle of hyphae, as in the true ectotrophics.

This broad spectrum of variation, from the endotrophic type through the many intermediates to the ectotrophic mycorrhiza of forest trees, indicates a high degree of tolerance on the part of the macro-symbionts towards mycorrhizal fungi. But if we regard the ectotrophic types as the more highly specialized, and the wholly endotrophic types as being little more than casual associations, a somewhat anomalous situation is revealed in mycorrhizal systems. Normally, one would expect increasing specialization in a symbiotic system to be correlated with a decreasing range of species participating in the associations. While this is perhaps true of the macro-symbionts, in the progression from endotrophic to ectotrophic association, it is far from true of the micro-symbionts. Endotrophic mycorrhizal fungi are, on the whole, limited to a few species only, including *Endogone*, *Rhizoctonia* and perhaps *Phoma* in the Ericales; but several hundred basidiomycetes participate in ectotrophic mycorrhizal associations.

3.4 Endophytic and endozoic algal systems

In addition to the lichen symbiosis, which has been treated separately, there are numerous symbiotic systems in which one of the symbionts is a blue-green alga. Significantly, in all known associations of this nature, the co-symbiont is either an autotrophic organism or is presumed to have been an autotrophe but to have lost its chlorophyll.

In a few liverworts, such as *Anthoceros*, *Blasia* and *Cavicularia*, species of *Nostoc* form cephalodia-like* structures on the ventral surface. They arise by migration of the *Nostoc* filaments into small cavities in the thallus which are later closed over by the growth of papillate cells. These associations have not been extensively investigated physiologically, but it has been shown by ^{15}N techniques that the *Blasia* symbiosis fixes molecular nitrogen (BOND and SCOTT, 1955) and that some of it is probably translocated to the macro-symbiont. Their carbohydrate metabolism still awaits investigation by ^{14}C methods. Until this has been done, we cannot say whether the micro-symbiont remains autotrophic in symbiosis, or whether it is partially heterotrophic.

A somewhat similar type of symbiosis occurs in the genus *Gunnera*. This is a plant with large radical leaves arising from a short rhizome. Slightly distal to the point of origin of each adventitious root on the rhizome of *Gunnera perpensa*, there appears a roughly triangular area of thin-walled cells of the cortex in which *Nostoc* filaments are to be found (Plate 3). These probably enter through the adventitious root primordia and migrate to the cortex of the rhizome.

Very little, if any, light penetrates the soil and leaf bases to the micro-symbiont. It is presumably heterotrophic for carbon, although even in deeply buried rhizomes the *Nostoc* remains dark green. Any contribution of fixed nitrogen from the *Nostoc* to the macro-symbiont must be negligible, considering the great diversity in the respective volumes of the symbionts.

The Cycad-*Nostoc* symbiosis reveals a close analogy to the legume symbiosis in the production of root 'nodules'. This group of plants is

---

*Gall-like structures on lichen thalli formed by association of the mycobiont with a secondary phycobiont.

characterized by the presence of coralloid roots near the soil surface which are usually interpreted as respiratory roots. Their coralloid nature is the result of repeated dichotomous branching.

Initiation of the symbiosis appears to result from migration of *Nostoc* filaments through ruptured root cells, into the middle cortex. Here they proliferate in and between the cells, and eventually form a complete ring round the stele (Plate 4). In some ways this is analogous to the construction of the lichen thallus. The location of the algal layer in cycad roots cannot, however, be determined by the indirect morphogenetic effect of light, as in the lichen symbiosis. There is little question of these being photosynthetic roots, although the *Nostoc* remains dark green even underground, as in the *Gunnera* symbiosis. As might be expected, cycad root nodules have a fairly high capacity for molecular nitrogen fixation (BOND, 1959).

These are only a few of the many examples of symbiosis between blue-green algae and autotrophic plants. There are equally many instances of symbiosis between other types of algae and green plants, such as *Chlorochytrium* in *Lemna* and other hydrophytes; *Phyllobium*, *Myxochloris* and *Chlamydomyxa* in *Sphagnum*, and *Phyllosiphon* in the leaves of Araceous plants in the tropics. The latter is one of the few instances, apart from the cycad symbiosis, in which any marked response on the part of the macro-symbiont is induced. In this case, gall-like structures are formed on the leaves.

Endozoic symbiotic systems, ranging throughout the Protozoa to the Coelenterates, have been the victims of diverse interpretation in the past. Just as in the lichen symbiosis, the phycobiont cells were formerly regarded as being either part of the invertebrate protoplasm or gametes. They are now, of course, known to be unicellular algae and are commonly termed Zooxanthellae, Cyanellae or Zoochlorellae according to whether their colour is yellow, blue-green or green. This arbitrary classification corresponds fairly closely to the Dinophyceae, Cyanophyceae and Chlorophyceae.

Although the number of symbiotic associations between invertebrates and unicellular algae is of the order of several hundred, it is a surprisingly small number considering the fact that marine invertebrates are in frequent ecological association with planktonic algae. Many such organisms ingest algae, whether or not they feed on them, so that there is a ready-made contact between potential symbionts. We must assume that the apparent 'resistance' to participation in symbiosis stems from the lack of physiological 'need' for symbiotic association on the part of the invertebrates. This is, to some extent, understandable because the marine environment is not characterized by sudden and local changes in nutrient status as is the terrestrial environment.

Despite the apparent evolutionary bias towards independent existence of aquatic organisms, however, it is here that we see some of the most specialized instances of symbiosis. The freshwater flagellate *Cyanophora paradoxa*, for example, lives in symbiosis with two or more blue-green algal cells which have been referred to *Cyanocyta korschikoffiana* (Plate 5). This symbiotic entity lives as an autotrophic organism. Reproduction of the two symbionts is simultaneous and the alga is considered to be incapable of free-state existence.

With several other freshwater protozoa, such as *Paulinella* and *Peliaina*, *Cyanophora* is very much on the borderline between the plant and animal kingdoms. Many other protozoa are also in this category, but they possess true chloroplasts (*Euglena, Chlamydomonas, Dinobryon*, etc.) although there is an element of doubt about this in some instances. The significant point is that here we have another instance of simulation, through symbiosis, of the autotrophic organism. *Cyanophora* 'utilizes' a blue-green alga as its chloroplasts and in this way it becomes as fully autotrophic as the euglenoids and others. If the *Cyanophora* phycobiont proves to have a nitrogen-fixing ability, in common with many other blue-green algae, this organism will be seen to have achieved an even higher degree of autotrophy than its evolutionary 'templates', just as blue-green lichens have surpassed the level of autotrophy of angiospermous plants.

3.5 Integration of multiple symbionts

The majority of symbiotic associations are effected between pairs of organisms, but the incidence of polysymbiotic systems is by no means rare. It is from these latter systems that we can perhaps gain a better understanding of the physiological interrelationships of symbionts. The lichen genus *Stereocaulon*, for example, clearly shows the progress of integration between more than two symbionts. The third symbiont of this genus, usually a species of *Nostoc, Scytonema* or *Stigonema*, is confined to structures termed cephalodia. In *Stereocaulon*, these are surface appendages formed by association of the blue-green alga with the mycobiont. They are of various shapes and sizes, and throughout the genus we can trace the gradual development of the cephalodium from the primitive type, in which the association is no more than a casual intermingling of secondary phycobiont filaments with mycobiont hyphae, to the highly differentiated nodular structure characteristic of *Stereocaulon ramulosum*. In this species, the secondary phycobiont has become completely enclosed in a spherical mycobiont structure with a short stalk and a cortex of isodiametric cells. Cephalodia of this type are practically confined to the pseudopodetia or 'stems' of the lichen (LAMB, 1951).

These highly differentiated cephalodia resemble in form, if not in function, the nitrogen-fixing root nodules of legumes. Tests for fixation of nitrogen by *S. ramulosum* cephalodia, using isotopic nitrogen, have shown that there is no significant uptake (SCOTT, unpublished data), but this does not imply that cephalodia of other species of *Stereocaulon*, containing a different secondary phycobiont, may not be active in this respect.

Stereocaulon cephalodia have presumably developed from the casual growth of free-living blue-green algae on and around the base of the plants. While no such series of development is evident in present-day root nodules, it is probable that a similar casual association was the starting point of root nodule development.

There are various records of the presence of more than one species of primary phycobiont in lichen thalli. While many of these have not been substantiated, it is clear that the occasion does arise when two species, or at least two physiological strains, occur in one and the same lichen species (see HALE, 1967, for review). It is not difficult to understand how different species of the same genus can become physiologically and morphologically integrated with the same mycobiont. Their basic physiology is very similar, and from what we now know of the exudation processes of lichen algae, it would appear that the taxonomic status of the phycobiont is not a potent factor in the observed morphogenetic effects in lichens.

The probability that all lichen symbioses are in reality multiple systems was given prominence some years ago by proponents of the theory that nitrogen-fixing bacteria constitute a third symbiont of the lichen system. No lichen, living under natural conditions, is free from epiphytic bacteria, so it is not surprising that the nitrogen-fixing bacterium *Azotobacter* has frequently been isolated from lichen thalli. But there is no evidence, from isotopic nitrogen tests on numerous lichens, of any significant amounts of nitrogen fixation that could be attributed to *Azotobacter* or other nitrogen-fixing bacterium.

Despite the lack of activity in the contribution of fixed nitrogen to the lichen symbiosis, the interesting question is raised as to whether the normal bacterial complement of lichens does in any way contribute to maintenance of the symbiosis. Lichen thalli, free from bacterial contamination, have in recent years been partially reconstituted from pure cultures of the symbionts (AHMADJIAN, 1966), but it would be premature at this stage to assert that lichen thalli can exist under natural conditions in the absence of a casual bacterial flora. Indeed, the more advanced our knowledge of symbiosis becomes, the more does it appear likely that such casual associations, as the bacterial 'contamination' of lichen thalli, contribute positively to the maintenance of symbiosis. This situation is, after all, strictly analogous to the situation among soil fungi and bacteria where it is now accepted that casual 'mutual aid' is the rule of the day in maintenance of soil populations.

Polysymbiosis in the field of mycorrhizal association is much more prevalent than in lichens. Although there are established instances of mycorrhizal association in which the relationship is obligatory, in the sense that the fungus does not produce fructifications when its connection with the tree is severed, for example *Amanita muscaria*, this failure has only been observed under particular edaphic conditions. In soils more congenial to free-living mycelial growth, there is no reason to suppose that fructification would not occur in the normal way.

Some mycorrhizal fungi take part in exclusive associations, for example, certain species of *Suillus* and *Lactarius* are confined to larch. Most trees, however, strike up an association with numerous mycorrhizal fungi depending upon locality, soil conditions and availability of the potential mycobionts. This can be readily verified by recording the occurrence of different species of agarics—many of which are mycorrhizal associates—growing in conifer plantations or deciduous woodland, under one tree species.

The very tolerant nature of most trees towards a variety of mycorrhizal fungi illustrates two noteworthy points of detail in which this type of symbiosis differs from the lichen symbiosis. First, the incidence of polysymbiosis throughout mycorrhizal systems is very high; second, it is of a much more casual nature than is the corresponding condition in the lichen symbiosis. At any time in the lifespan of a single tree, the initial mycorrhizal associate may be superseded by another species forming an association on later-formed roots. This replacement of one symbiont by another has never been observed in any lichen species.

There is no evidence that multi-mycorrhizal systems of the ecotrophic type on the same tree are in any way cyclically interdependent. This is more a case of several entirely separate systems, in the physiological sense, than is the presence of cephalodia in or on lichen thalli. Since the secondary phycobiont in the majority of cephalodial lichens is a species of *Nostoc*, cyclic interdependence is more obvious. *Nostoc* is a nitrogen-fixing alga and can therefore be presumed to supply fixed nitrogen to each of its co-symbionts.

Recent work by FOSTER and MARKS (1967) points to the simultaneous association of up to seven forms of mycorrhizal fungi with *Pinus radiata* (Monterey pine) in Australian forest soils. They have shown these forms to be morphologically distinct and they have also shown that the bacterial 'shell' associated with the mycorrhizosphere (the equivalent of the rhizosphere in normal roots) can be specifically different for the various mycorrhizal types. It is evident that some of these bacterial associates are nitrogen fixers and that they are associated with a particular region of the mycorrhizosphere which includes the outermost layers of the mantle. Senescent hyphae predominate here and it is assumed by Foster and Marks that the bacteria derive their carbon nutrition from these hyphae, possibly in the form of mannitol.

Although this nitrogen-fixing mycorrhizal association does not fall strictly within the confines of a polysymbiotic system, in the sense used here, because the bacteria are 'distant' associates, or at best are present as casual surface associates of the fungal hyphae, the system functions physiologically as such. By extension, it is probable that mycorrhizal systems of other trees will be shown to be similarly organized as polysymbiotic systems, thereby finally closing the long chapter of controversy over the nitrogen-fixing ability of mycorrhizal fungi.

How far the principle of polysymbiosis applies to endotrophic mycorrhizal systems is not so well known. In the vesicular-arbuscular types, the mycobionts appear to be exclusively species of *Rhizophagus* or *Endogone*, but there is apparently more liberalism in other endotrophic systems.

One of the clearest examples of polysymbiosis that has recently been investigated by modern experimental methods is the triple symbiotic system involving a tree, a mycorrhizal fungus and the achlorophyllous plant *Monotropa* (§3.3 and Fig. 4–5). The important point about this triple system is that it illustrates the ability of a single mycorrhizal fungus to effect concurrent associations with two widely different plant species.

This is not a unique occurrence, for the parasitic fungus *Armillaria mellea*, which is responsible for so much damage to a wide variety of tree species, also associates symbiotically with the orchid *Gastrodia elata*. The *Armillaria-Gastrodia* complex is really a phase-separation symbiosis. It is symbiotic in the sense that there is bilateral movement of carbohydrates between the partners, but this is separated in time. This partnership is usually described as one of initial parasitism of the orchid by the fungus, followed by parasitism of the fungus by the orchid. The tuber of the orchid is said to remain dormant until it is parasitized by *Armillaria*, whereupon it is stimulated in some unknown way to produce a flowering scape. Energy for scape formation is thought to be derived from carbohydrate of fungal origin, thus constituting a reversal of the initial direction of carbon flow and a reversal of the parasitic relationship. In effect this amounts to a phase-separation symbiosis.

Symbiotic organisms are remarkably tolerant of each other in respect of their ability to associate with more than one co-symbiont either concurrently or consecutively. Despite this, every symbiotic association is attended by profound changes in morphology of one or more of the symbionts. How much these effects are physical or biochemical, we are as yet largely unaware. Further research into the whole problem of morphogenesis in symbiotic systems is urgently needed, not only to clarify the intricacies of symbiosis *per se*, but to help open up the much wider field of morphogenetic effects in plants and plant organs in general.

Physiological Integration 4

4.1 Interflow of metabolites

Perfect symbiosis could be defined as the condition of such intimate association between two organisms that neither can complete its vegetative or reproductive life cycle in the absence of the other. Intrinsic to this ultimate condition is absolute nutritional interdependence. We can recognize various stages in the progression towards this ideal of coexistence but there is no known example of its complete achievement. The nearest approach is perhaps the *Cyanophora paradoxa* or the *Paulinella chromatophora* symbiosis (§3.4).

Symbiotic association is a nutritional alliance induced by the lack of one or more physiological attributes associated with nutritional self-sufficiency. Fungi lack chlorophyll, but some have formed the lichen symbiosis in which the fungus gains the benefits of an autotrophic partner and in the process achieves a remarkable simulation of the angiosperm leaf. Others have resorted to association with chlorophyllous vascular plants. Numerous invertebrates have formed associations with unicellular algae, some of them, such as certain anemones, bivalves and corals, acquiring the additional attribute of photosensitivity.

All known plant symbioses are initiated and maintained by the unilateral or bilateral movement of metabolites. Chief of these is organic carbon, but other nutritional and physical factors act as supplements to coexistence of the symbionts. The simplest and the most sophisticated symbioses have a common link in the interflow of metabolites—the process to which we must look for the *raison d'être* of all symbiotic life.

4.1.1 Water and inorganic nutrients

Plant symbioses can be categorized on the basis of their water and nutrient relations. In one group there is total dependence of one of the symbionts upon the other for water and inorganic nutrient supply. Most of the lichen and endozoic symbioses, root- and leaf-nodule and blue-green algal symbioses are included here. Dependence is gained by removal of one of the symbionts from the sphere of influence of the substrate by enclosure within the tissues of the co-symbiont (e.g. *Rhizobium* in root nodules), or by physical elevation from the substrate (e.g. the secondary phycobiont of *Stereocaulon*). In the other group, the symbionts are partly or wholly independent. Many Angiosperms and Gymnosperms, for example, show partial dependence on their mycorrhizal symbionts for nutrient uptake.

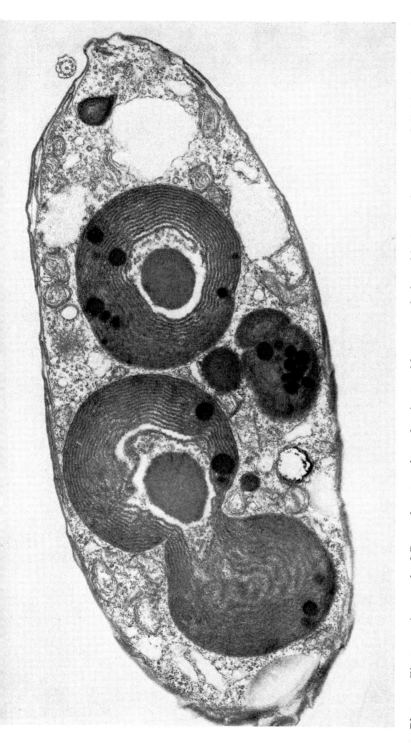

Plate 5 Electron micrograph of *Cyanophora paradoxa*, showing two blue-green algal symbionts, one in process of division. (×21,000) (Courtesy of William T. Hall)

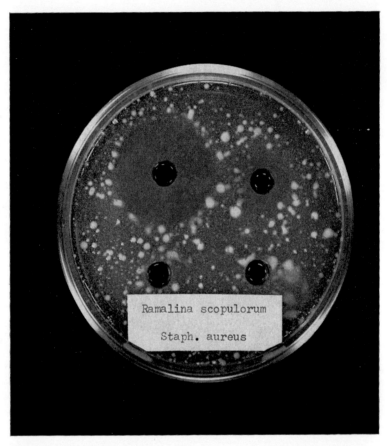

Plate 6 Antibiotic activity of usnic acid from the lichen *Ramalina scopulorum*.

Symbiotic systems are peculiarly susceptible to changes in the nutrient status of either of the symbionts. This is well illustrated by the lichen symbiosis. It has been shown, for example, that when certain species are cultured under laboratory conditions the maintenance of symbiosis between alga and fungus is dependent to a considerable extent upon the supply of inorganic nutrients (SCOTT, 1960). On the one hand, excessive nutrient supply has the effect of causing the breakdown of the symbiotic union (Fig. 4–1); on the other hand, deficiency in general nutrient level has not been shown to have any deleterious effect on the symbiosis, apart from considerably lowering the rate of symbiotic growth.

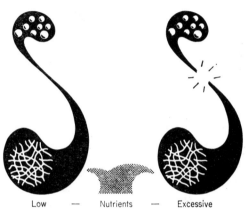

Low — Nutrients — Excessive

Fig. 4–1 Nutrient excess, under non-limiting conditions of light intensity and moisture, causes breakdown of the lichen symbiosis. The letter **S**, symbolic of symbiosis, displays in diagrammatic form the relationship between symbiotic organisms. The small hook represents the micro-symbiont in permanent though frequently tenuous contact with the macro-symbiont, represented by the large hook.

Numerous attempts have been made from time to time to reconstitute a lichen thallus from the isolated symbionts, but little success was achieved until it was realized that deficiency in general nutrient level was an important factor in the 'forcing together' of the prospective symbionts. In this case there is an apparent link between growth rate and photosynthesis of the phycobiont on the one hand, and growth rate of the mycobiont on the other. Excess results in overgrowth of the phycobiont and breakdown of the symbiosis.

The initiation of ectotrophic mycorrhizal associations with forest trees has been shown to be susceptible to the nutrient status of the soil and the host plant. When this is high and balanced the incidence of symbiotic initiation is low, but when it is at a low level the necessary conditions are created for establishment of the association (Fig. 4–2). This is but one of the several factors which may be co- or contra-operative in mycorrhizal initiation. That there are others, particularly edaphic or microbiological, is indicated by instances of better mycorrhizal association in loam soils than in nutrient deficient soils.

Apart from their importance in the initiation of mycorrhizal associations, inorganic nutrients are concerned in the functional symbiosis between tree and fungus. It has been assumed for many years that nutrient uptake by mycorrhizal roots surpasses that of asymbiotic roots. Much of the evidence for this assumption arose from observation of forest stands on nutrient-deficient soils. More substantial evidence has been obtained by feeding

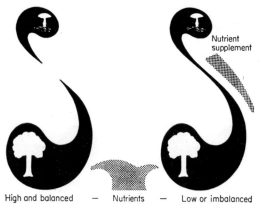

Fig. 4-2 Mycorrhizal initiation is impeded by high and balanced soil nutrient status, but is stimulated by low or imbalanced status.

mycorrhizal and non-mycorrhizal roots with radioactive tracer elements such as phosphorus, calcium and potassium (see review by MEYER, 1966). This function of mycorrhiza is clearly not an obligate one, but instances have been recorded which illustrate that survival of the tree can be decisively dependent upon nutrient uptake by the mycorrhizal system.

The general pattern that now emerges is a system of nutrient uptake from the soil by the mycorrhizal fungus at the expense of respiratory energy. There is evidence that this uptake is associated with accumulation of nutrients by the fungal mantle surrounding the roots (HARLEY, 1959).

The very common vesicular-arbuscular type of endotrophic mycorrhizal association (§3.3) is a further example of the promotive effect of nutrient deficiency on symbiosis initiation. Much of the work on synthesis indicates that phosphorus deficiency in particular leads to successful initiation. This has been demonstrated for tobacco, tomato and maize by DAFT and NICOLSON (1966). Once established, the vesicular-arbuscular mycorrhiza is apparently similar in function to ectotrophic systems in that nutrient uptake by the host plant is stimulated. This function is particularly important in species such as *Agathis australis*, which have become adapted to growth in nutrient-deficient soils.

These instances of correlation between nutrient status and symbiotic association reveal symbiosis as a physiological makeshift. Deficiency of nutrients is inimical to plant growth, yet this very condition stimulates the assumption of the symbiotic state. Once this is achieved, in mycorrhizal systems at least, the synergistic* effect on nutrient balance overcomes the deficiency apparent in the asymbiotic condition.

4.1.2 Carbohydrates and other metabolites

The keystone of symbiotic association is the sharing of organic carbon between the symbionts. In the majority of cases this results from the saprophytic activity of a non-green symbiont.

Several symbiotic systems have been investigated for movement of carbohydrate between the symbionts, using the radioactive isotope ^{14}C. This has proved to be a particularly useful tool in studies which, perforce, involve the tracing of organic carbon from one organism to another.

The pattern of carbohydrate translocation in the lichen genus *Lobaria* has recently been established (RICHARDSON et al., 1967) by illuminating discs of the lichen thalli on media containing labelled bicarbonate as a source of CO_2. Two of the species used, *Lobaria pulmonaria* and *L. laetevirens*, have *Trebouxia* (Chlorophyceae) phycobionts, while the third species, *Lobaria scrobiculata*, has a *Nostoc* phycobiont. In all three species the translocate eventually appeared in the tissues of the mycobiont in the form of mannitol, but the intermediate carbohydrates differed according to the type of phycobiont. In the blue-green *Lobaria*, the evidence indicates that glucose is exuded by the phycobiont (*Nostoc*) and converted to mannitol by the mycobiont. In the green species, however, it appears that the translocate is the sugar alcohol ribitol. This pentose derivative has also been isolated from *Xanthoria* and has been detected as an exudation product of the phycobiont (*Trebouxia*) isolated from the lichen.

Experiments with cultured discs of the lichen *Peltigera praetextata* (SCOTT, 1960) have shown that the growth rate of the holobiont is directly related to light intensity and that growth can be adequately supported for periods of several months on a nitrogen-free inorganic medium. Under laboratory conditions, therefore, it is evident that carbohydrate, elaborated by the phycobiont, is transferred to the *Peltigera* mycobiont in sufficient quantity to provide its total carbon nutrition.

To what extent the one symbiont of *Peltigera* is dependent upon the other for carbohydrates under natural conditions is another matter. The important point has been established, however, that a lichen mycobiont *can* be totally dependent upon its chlorophyllous co-symbiont.

---

*Synergism = a co-operative effect produced by two organisms that is greater than the sum of the effects produced by the organisms growing separately.

Evidence arising from recent attempts to induce the artificial synthesis of lichens (AHMADJIAN, 1966) indicates that one of the requirements for successful association of alga and fungus is the restriction of carbon nutrition to that derived from the prospective algal symbiont. The presence of organic nutrients in the synthesis medium prevents association taking place. This is perhaps understandable because preferential utilization of an exogenous source of organic nutrition eliminates the physiological 'need' for initiation of symbiosis (Fig. 4–3).

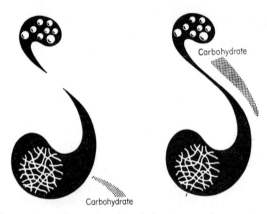

Fig. 4–3 An exogenous source of carbohydrate prevents initiation of the lichen symbiosis. Only when there is a physiological 'need' for symbiosis does successful initiation occur under the influence of carbohydrate and other metabolites exuded by the phycobiont.

Carbon nutrition studies of numerous lichen mycobionts in pure culture indicate strong preferences for simple carbohydrates, although many can also use sugar alcohols. Starch and cellulose are not easily utilizable carbon substrates. Preference for the simpler carbohydrates is a characteristic physiological feature of parasitic fungi, and the degree to which they can also utilize higher polysaccharides and lignin is a fairly reliable measure of their ability to live as saprophytes.

Lichen mycobionts appear to conform to the pattern of parasitic fungi and it is therefore doubtful whether they have any capacity for asymbiotic existence. This conclusion is supported by the fact that no substantiated reports of free-living mycobionts have ever been published.

Radioactive carbon studies of ectotrophic mycorrhizal systems have provided the final proof that photosynthate is transferred from the chlorophyllous macro-symbiont to the mycorrhizal fungus. MELIN and NILSSON (1957) demonstrated this by feeding labelled CO_2 to pine seedlings and later isolating labelled carbon compounds from the fungal sheath surrounding the mycorrhizal roots. The basic concept of symbiosis has thus been proved for ectotrophic mycorrhizal systems. As to the necessity of an endogenous source of carbohydrate for survival of the symbiosis, there is little information. There have been convincing demonstrations of the inability of certain mycorrhizal fungi to produce fructifications following severance of their mycelial connections with the tree. While this proves the point regarding adherence to the concept of symbiosis, it does not prove that the mycelium is incapable of free growth in the soil.

The majority of mycorrhizal fungi isolated into pure culture can only utilize mono- and disaccharides, although it is evident that they are nutritionally a very heterogeneous group. Several important mycorrhizal associates can digest cellulose and have been shown to have a high cellulolytic activity. But we cannot assume that there is preferential utilization of cellulose by these fungi in symbiosis.

The similarities in carbon nutrition between mycorrhizal and lichen fungi are not without significance. Both groups are generally unable to utilize higher carbohydrates or lignin but there are exceptions, particularly amongst the mycorrhizal fungi. For the exceptions, symbiosis versus free-living saprophytism is evidently a function of substrate versus potential co-symbiont influence.

There is a well-established relationship between light intensity and the initiation of mycorrhizal symbiosis. Extensive work on this aspect has been done in Sweden by BJÖRKMAN (1949) who has shown that the percentage of mycorrhizal root tips in pine seedlings increases in direct relationship to increase in light intensity. The most interesting feature of Björkman's results is the appearance of a saturation light intensity above which little further increase in mycorrhizal association takes place. This indicates a correlation between some photosynthetic product and mycorrhiza formation. KINUGAWA (1965) has shown that the number of mycorrhizal roots formed by seedlings of *Pinus densiflora* under continuous light is greater than under light-break treatment or under short-day conditions. This work eliminates the existence of a photo-periodic effect in mycorrhiza formation.

It has been suggested that the effects of light and nutrient status on the initiation of mycorrhizal associations are linked through the carbohydrate level of the macro-symbiont roots. A high level may favour the growth of mycorrhizal hyphae from associations that have already taken place and may stimulate the process of association itself as the result of exudation of metabolites from the roots into the rhizosphere (Fig. 4–4). Since many mycorrhizal fungi have little ability to utilize either cellulose or lignin, they will respond to the presence of glucose or other simple carbohydrate in the root exudate of a potential macro-symbiont.

Numerous apparent contradictions of the carbohydrate theory have been demonstrated, such as the stimulation of mycorrhizal association by the *addition* of nutrients to soils already of a high nutrient status. Further, with the demonstration that mycorrhizal fungi, such as species of *Amanita*, *Boletus* and *Coprinus*, can produce IAA (ULRICH, 1960), it has become evident that some of the effects hitherto ascribed to nutrient status and light intensity might be explained by the effect of auxin secretion by the mycorrhizal fungi in the soil or in the macro-symbiont roots.

Fig. 4-4 Mycorrhizal initiation is correlated with light intensity. The link in this relationship may be the carbohydrate status of the macro-symbiont roots.

Indole acetic acid is known to stimulate the translocation of carbohydrate within plants and thus auxin production by mycorrhizal fungi may account for at least a proportion of the high carbohydrate level of mycorrhizal roots. Whether root carbohydrate is to be regarded as a factor in mycorrhizal initiation, or as the auxin-induced result of association, remains to be seen.

The 'saprophytic' plant *Monotropa* (§3.3) usually grows in the vicinity of conifers or broad-leaved trees such as beech and oak, and was for many years considered to be a forest litter saprophyte by virtue of its mycorrhizal association. This is substantially true, but BJÖRKMAN (1960) has confirmed experimentally that the mycorrhizal association extends to the forest trees. Labelled glucose was injected to the phloem of *Pinus* and *Picea* trees growing in the vicinity of *Monotropa* plants. Four days later, significantly greater ^{14}C activity was found in *Monotropa* plants within 1·6 metres of the injected trees than in control plants at distances of more than 17·5 metres from the trees. Injection of ^{32}P-phosphate to the tree phloem resulted in similar translocation. FURMAN (1966) gives additional evidence of the identity of the *Monotropa* mycorrhiza with that of trees. In this instance, ^{32}P was shown to be translocated in the reverse direction from *Monotropa* stems to the ectotrophic mycorrhiza of *Quercus* and to the endotrophic mycorrhiza of *Acer*.

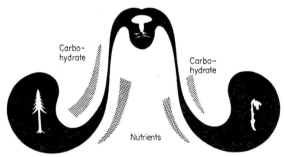

Fig. 4-5 The Conifer-*Boletus*-*Monotropa* association. Carbohydrate is channelled from the tree leaves through the common mycorrhizal fungus to *Monotropa*.

The *Monotropa* symbiosis is thus a three-tier system involving a chlorophyllous symbiont, a non-chlorophyllous symbiont and a fungal bridge connecting the two (Fig. 4-5). If further proof were needed of the functioning of this system it is to be found in Björkman's synthesis of an ectotrophic mycorrhizal association between both *Pinus* and *Picea* seedlings and the mycobiont (? *Boletus*) isolated from *Monotropa*.

The remarkable feature of this multiple association is the ability of the mycobionts to act dual roles. In ectotrophic association with forest trees, the normal intercellular Hartig net and extradermal mantle are formed; but in association with *Monotropa* these characteristics are supplemented by the formation of haustoria which penetrate the epidermal cells. The endotrophic symbiont of the *Acer-Monotropa* complex is apparently more consistent in its relationship with the two species. In each case the mycobiont reveals antithesis of physiological function in its association with tree and *Monotropa*—on the one hand, extraction of carbohydrates from the chlorophyllous symbiont; on the other, contribution of carbohydrate to the non-chlorophyllous symbiont.

The presence of haustoria in the epidermal cells of *Monotropa*, even when it is aligned with an ectotrophic mycorrhizal system, implies that the fungus has an active cellulolytic enzyme complex which is antagonized only to the extent that the haustoria are confined to the cells of the epidermis. In the tissues of its alternative symbiont, however, this enzyme activity is evidently inhibited by a defence mechanism of the root cells.

Numerous other saprophytic plants, for example *Neottia* and *Corallorhiza*, are entirely dependent on their mycorrhizal associates for carbon nutrition. Whether they too take part in a multiple symbiosis of the

Monotropa type is uncertain, although possible, in view of Furman's work on the endotrophic mycorrhizal connection between *Acer* and *Monotropa*. The mycorrhiza of *Neottia* and *Corallorhiza* (Plate 2), in common with other orchids, is of the endotrophic type and the micro-symbionts have been identified as species of *Rhizoctonia*. If endotrophism is compatible with the *Acer-Monotropa* symbiosis, there is reason for anticipating the demonstration of a link between the micro-symbionts of saprophytic orchids and a third symbiont.

The relationship between *Rhizoctonia* and orchids, obligate only in the sense that asymbiotic seed germination and early development are greatly impaired if not entirely prevented, is one which is delicately but imprecisely poised between parasitism and symbiosis. Observations of the symbiosis under both cultural and field conditions have revealed that orchid seedlings are frequently associated with more than one micro-symbiont. In pure culture the latter can utilize a wide range of carbohydrates including cellulose. Some can also break down lignin.

Substrate diversity is illustrated by the occurrence of *Rhizoctonia solani* as a micro-symbiont of the green orchid *Dactylorchis purpurella* (DOWNIE, 1959) and by the occurrence of its perfect, or sexual stage, *Corticium solani*, as an important pathogen of crop plants including potato and sugar beet. Here we have an example of metabolic adaptation to substrate. As *Corticium solani*, the typical parasitic habit of carbohydrate removal from the host is revealed. As *Rhizoctonia solani*, carbohydrate is contributed to the orchid seedling.

A second micro-symbiont of *Dactylorchis*, identified as *Rhizoctonia repens*,* has been investigated for carbon nutrition by SMITH (1966). Translocation of glucose from fungus to orchid was demonstrated by ^{14}C techniques. There is also some slight indication that glucose uptake by the orchid is stimulated when cellulose is made available to the fungus.

It is not known with certainty that there is any physiological advantage to the fungus resulting from its participation in the orchid symbiosis. As soon as the seedling orchid becomes established as a photosynthetic unit, the nutritional requirement for symbiosis falls away, at least in those orchids which become fully autotrophic in their later growth. If there is loss of symbiotic equilibrium, with reversion of the micro-symbiont to parasitism, the association retains its nutritional validity. This constitutes a special case of association which can be considered as a phase-separation symbiosis (Fig. 4-6). One symbiont is nutritionally sustained until it becomes autotrophic; the other becomes dependent, at least partially, on the now autotrophic symbiont by virtue of the reversal of the carbohydrate diffusion gradient.

The orchid symbiosis represents the borderline between mutual tolerance and antagonism. These two conditions can pertain within the lifespan

---

*Now recognized as the imperfect stage of the basidiomycete *Tulasnella calospora*.

of the orchid. The percentage of seedlings attaining physiological equilibrium with the micro-symbiont is relatively small under natural conditions; of those that *do* attain this equilibrium, only a few maintain it until photosynthesis can supply the total energy requirement.

The enigmatic feature of this symbiosis, which still awaits clarification, is the apparent capacity for reversal of carbon flow according to substrate conditions and the physiological vigour of the orchid.

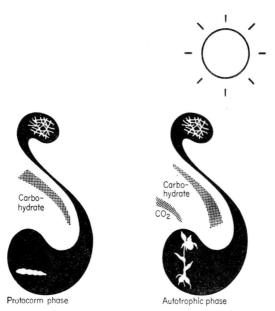

Fig. 4–6 The phase-separation symbiosis in orchids. Germination and the protocorm phase are supported by carbohydrate from the mycorrhizal fungus. During the autotrophic phase of the orchid, carbohydrate is transferred to the mycorrhizal fungus.

By analogy with the initiation of mycorrhizal associations, it is to be expected that carbohydrates will have some effect on the nodulation of leguminous and other plants. Carbohydrate influence on nodule initiation is thought to be correlated with the combined nitrogen status of the macro-symbiont. Thus, when a high level of carbohydrate is induced in the plant by increase in the rate or duration of photosynthesis, or by supplying an exogenous source of carbohydrate, the inhibitory effect of combined nitrogen on nodule initiation is substantially reduced.

The carbohydrate status of the macro-symbiont may well be an inductive factor of nodule initiation, as it may also be for mycorrhiza formation. But, as in the mycorrhizal association, it may also be the result, not the cause, of entry of the micro-symbiont: and for the same reason, that the formation of IAA is induced within the root tissues and so stimulates carbohydrate translocation to the roots.

The rhizosphere carbohydrate effect is illustrated by the numerous studies on the effects of exogenously applied carbohydrate on nodulation. It is apparent that sugars added to the growth medium have, under certain circumstances, a generally depressant effect on nodulation, in the relatively high concentrations that have been used. This should be expected, since added carbohydrate will be evenly distributed in solution throughout the medium and will therefore promote asymbiotic growth of the prospective micro-symbiont.

In contrast to these experimental conditions, exudation of carbohydrates from the roots into the rhizosphere sets up a concentration gradient from root cells to the outer limits of the rhizosphere. The stimulus is thus provided for the micro-symbiont to 'grow up' the concentration gradient towards the root hairs where penetration occurs.

There is ample evidence in rhizosphere literature of the selective effect of root exudates on the growth of soil micro-organisms, particularly those that can only use simple carbohydrates and those that are heterotrophic for certain amino acids or vitamins (THIMANN, 1963). Thus, the participation of carbohydrate as an inducement factor to association is only part of the complex, but an obviously important one.

Compared with the nutritional relationships that have evolved between terrestrial plants, symbiosis between unicellular algae and aquatic invertebrates does not appear to have resulted in the circumvention of natural carbon food chains. This should not be unexpected, for in every case the zoobiont (symbiotic animal) is capable of movement, however limited, in a more uniform medium than that of terrestrial organisms. Active ingestion by both carnivorous and herbivorous zoobionts implies a certain element of choice and therefore a low evolutionary pressure towards the selection of individuals physiologically adapted to an alternative source of carbon nutrition. This is in direct contrast to all plant symbionts for which intake is dependent upon concentration gradients of one kind or another.

Few of the many zoobionts—numbering more than a hundred genera—have become specialized to the extent of total physiological dependence upon their co-symbionts. Some, for example the marine worm *Convoluta roscoffensis*, the freshwater flagellate *Cyanophora paradoxa* and the rhizopod *Paulinella chromatophora*, have achieved this condition insofar as their carbon and probably also their nitrogen nutrition is concerned. *Cyanophora* and *Paulinella* perhaps represent the nearest approaches to what I have called the perfect symbiosis (§4.1).

In *Cyanophora* particularly, the work of HALL and CLAUS (1963) has shown that there is considerable modification of the morphological structure of the endozoic alga (§3.4). The cell wall has apparently been dispensed with and there is a very close approach, in function at least, to the chloroplast of the green plant cell.

It can scarcely be doubted that the phycobionts of symbioses with this grade of specialization contribute all the carbon materials required by the holobiont. *Paulinella*, for example, which contains two phycobiont cells, has never been observed to ingest food particles.

The situation is not so clear in many other endozoic symbioses. From the relatively few published works on carbon nutrition using ^{14}C techniques, it is apparent that at least some symbiotic systems reveal the transference of carbohydrate or other carbon-containing substances from the phycobiont to the zoobiont. MUSCATINE and HAND (1958) have shown that the phycobiont of the sea anemone, *Anthopleura elegantissima*, takes up ^{14}C when illuminated in sea-water containing labelled bicarbonate, and that ^{14}C is translocated in some form to the other epithelium of the zoobiont. This tissue is normally free of phycobiont cells.

More recently, MUSCATINE and LENHOFF (1963) have demonstrated incorporation of ^{14}C by the zoobiont of *Chlorohydra viridissima* from photosynthesis by the phycobiont. None of these studies, however, indicates to what extent incorporation of carbon from the phycobiont is essential to maintenance of the symbiosis. A start in this direction has been made by GOREAU and GOREAU (1960) who have shown that the uptake and distribution of ^{14}C in two reef-building corals, *Manicina areolata* and *Montastrea annularis*, is not such that the zoobionts could at any time be entirely dependent upon carbohydrate photosynthesized by their phycobionts.

Although the majority of coral-forming species are symbiotic, it is apparent that this is one instance of symbiosis which has shown little progression towards autotrophism. At the present stage of its evolution we can discern only a short-circuiting of a natural food chain on the part of the phycobiont; that is to say, the products of zoobiont excretion are utilized directly by the phycobiont as an alternative to absorption from the seawater (§4.3.2).

4.2 Unique physiological processes

Physiological association between plants is characterized by the occurrence of biosynthetic pathways functioning in response to stimuli resulting from the act of association. In the host-parasite relationship, specific metabolites produced by the parasite create a stimulus to which the host responds by the elaboration of 'antibodies'. The response constitutes a defensive reaction, mediated by adaptive enzyme systems, which creates a condition of incompatibility between host and parasite.

Development of species association to the symbiotic or mutual tolerance level has been accompanied by the elaboration of physiological processes which, although they may have been evolved from defensive reaction systems, are now recognized as being processes that make a contribution to the association. Some of these are unique because they are of extracellular origin; they function in the maintenance of the association and they are not known to be characteristic of either symbiont in free growth. To this latter extent there is a similarity to the defence reaction mechanism in the host-parasite relationship, but here the similarity ends. The defence reaction is a metabolic pathway, stimulated by an exotic metabolite, whose end-point is a cell substance or condition inimical to the parasite. The unique symbiotic process is likewise a train of metabolic events stimulated by the association, but whose end-point is a specific metabolite that participates in the maintenance of the symbiotic state.

Two important processes of this nature are the metabolism of lichen acids induced by the association of algae and fungi, and the fixation of molecular nitrogen induced by the association of vascular plants with the bacterium *Rhizobium* or an organism resembling an actinomycete.

4.2.1 Lichen acid metabolism

Lichen acids are probably produced by all except the gelatinous lichens and a few other isolated species. In most cases the distinctive colouring of lichens is due to the presence of one or more of these acids. If practically any coloured lichen thallus is sectioned and examined under a microscope, crystals of lichen acids will be seen adhering to the mycobiont hyphae, particularly on the surface of the cortex and on the hymenial layer of the apothecia. They are easily extracted with acetone and can be crystallized from a variety of solutions (HALE, 1967).

Chemically, lichen acids are mostly aliphatic and aromatic compounds. The majority of the known acids are either depsides or depsidones (condensation products of phenolic acids) and include such well-known examples as evernic, gyrophoric, lecanoric, lobaric, physodic, psoromic, stictic, thamnolic and usnic acids. About a hundred lichen acids have now been fully investigated and their chemical structure determined.

There are few authenticated records of any lichen fungus or alga being able to synthesize lichen acids in pure culture. Numerous phenol carboxylic acids, presumed to be precursors in lichen acid synthesis, have been identified in cultures of lichen mycobionts (HESS, 1959). The work of Hess and others indicates that lichen mycobionts carry the synthesis of lichen depsides and depsidones as far as orsellinic acid or related compounds. Orsellinic acid (methyl-dihydroxy-benzoic acid) is thought to be derived by the condensation of acetyl CoA (co-enzyme A) with malonyl CoA. The example cited by MOSBACH (1964) postulates the condensation of one molecule of acetyl CoA with three of malonyl CoA to give three molecules of orsellinic acid. These suggested precursors of orsellinic acid are readily available in plant cells. Acetyl CoA, at least, is well known as an intermediate in the entry of pyruvate to the Krebs cycle, and a similar system is thought to operate in the formation of pigments in free-living microorganisms.

MOSBACH and EHRENSVÄRD (1966) have shown that cell-free extracts of *Umbilicaria pustulata*, and of the phycobiont of a related species, *U.*

§ 4.2 UNIQUE PHYSIOLOGICAL PROCESSES 41

papulosa, can hydrolyse the tridepside gyrophoric acid to orsellinic acid. The enzyme system, presumed to be an esterase, is apparently specific to gyrophoric acid and two others, evernic and umbilicaric acids. Another enzyme in the extract of the holobiont, but not in the phycobiont extract, and therefore presumably of mycobiont origin, was shown to decarboxylate orsellinic acid and certain of its derivatives to their constituent phenols. The proposed scheme for degradation of depsides is thus:

Depside $\xrightarrow{\text{esterase (phycobiont)}}$ phenol carboxylic acids (e.g. orsellinic acid)

Phenol carboxylic acids $\xrightarrow{\text{decarboxylase (mycobiont)}}$ phenols (e.g. orcinol)

The lack of orsellinic acid decarboxylase activity in the phycobiont extract suggests that this enzyme is of mycobiont origin, but pure cultures of the *Umbilicaria* mycobiont have yet to be examined for esterase or decarboxylase activity.

This demonstration of depside hydrolysis by lichen extracts cannot be taken as evidence that synthesis follows the reverse pathway. It does

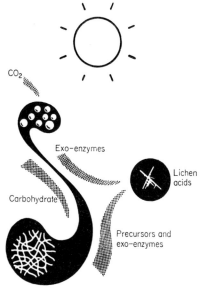

Fig. 4–7 Lichen acids are formed on the surface of the cortical hyphae by an exo-enzyme and precursor system which may originate from both symbionts.

indicate, however, that enzyme systems of both symbionts may be involved in the process (Fig. 4–7). We do not yet know what these enzymes are, nor at what points they participate in the synthesis.

Lichen acid formation is correlated to light intensity and duration, as might be expected from the known carbon metabolism of the holobiont. But numerous observations indicate that other factors are involved. There is, for example, the instance of the production of physcion (parietin) in the apothecia of *Xanthoria parietina* growing in deep shade (SCOTT, 1964). The hymenium, which contains no phycobiont cells, appears deep orange in colour while the rest of the thallus is quite green and almost devoid of physcion. Evidently there is a local accumulation of physcion precursors or of the enzyme system on the paraphyses of the apothecium, for it is here that the physcion is deposited. This is independent of the direct effect of light on phycobiont photosynthesis and suggests the intervention of internal factors which are a function of the organography of the plant.

Many lichen acids have been shown to have antibiotic properties (Plate 6). The majority of those investigated are active against Gram-positive bacteria and a few, notably usnic acid, inhibit the growth of certain pathogenic organisms. Anti-bacterial activity of lichen acids may be a protective mechanism for the symbiosis against invasion of bacteria which would cause disarrangement of the symbiotic balance by their metabolic effects. Most soil bacteria, however, are Gram-negative and are therefore unaffected by lichen acids.

The antibiotic activity of lichen acids can be demonstrated by the Oxford cylinder technique. Acetone extracts of the lichens should be diluted with water to about 40% acetone concentration and autoclaved. Pipette into wells cut from nutrient agar plates which have been seeded with a Gram-positive bacterium suspension before pouring. The lowest effective concentration is determined by making serial dilutions of the extract with 40% acetone. After incubation, the diameter of the clear ring round each well gives a measure of the extent of growth inhibition.

Several other ways in which lichen acids may participate in the symbiosis have been suggested from time to time. They have been thought to serve in a defensive capacity against attack by slugs and other small animals, due to the bitter taste imparted to the lichen thallus. There is little evidence to support this view and indeed, the giant millipedes of the tropics are known to have voracious appetites for highly pigmented lichens such as *Dermatiscum* and *Acarospora*.

Recent suggestions are that they may be concerned in the translocatory processes between the symbionts and that they act as chelating (metal binding) agents which function in the removal of minerals from rock substrates. Participation in translocation has been suggested, not on the basis of direct evidence, but in the light of the ability of some lichen acids to increase the permeability of animal membranes. By extension to plant cell membranes, it is supposed that the phycobiont cells are rendered more permeable to carbohydrates, thus creating the conditions for rapid exudation rates. The ability of lichen acids to bind metal ions from rock substrates has been known for some time and in this facility there is a possible advantage to the symbiosis. It cannot, however, be regarded as a function of the symbiotic unit itself, but as an incidental, though perhaps useful, extraneous feature.

None of these possible functions of lichen acids can be considered as essential factors in the continued association of the symbionts. Their great variety of production alone militates against this, as does the fact that many species comprise several phytogeographic races containing different combinations of a common pool of lichen acids. We can, however, assign these functions to assistance in maintaining the observed ecological amplitude of the individual species. In this we can also include the probable function of coloured lichen acids in stabilizing the amount of light reaching the phycobiont layer (§4.3.1).

4.2.2 *Molecular nitrogen fixation*

The symbiotic associations between *Rhizobium* and leguminous plants, and between actinomycetes and non-leguminous plants, have a common property in the reduction of molecular nitrogen in the root nodules (§3.2). Such reduction is sufficient to enable the symbionts to complete their vegetative and reproductive growth with no external source of combined nitrogen.

In legumes, the relationship between the symbionts conforms to the pattern of all incompletely integrated symbiotic systems in that they are separated by membranes, whether or not the one symbiont is habitually present 'inside' the cells of the other (Chapter 3). Since neither of the symbionts is known to reduce gaseous nitrogen asymbiotically, it is probable that the symbiotic process takes place on or between the membranes, as does the formation of lichen acids in the lichen symbiosis (§4.2.1).

Attempts to localize the site of nitrogen fixation in the nodules have provided data that can be variously interpreted. Experimental procedures have so far been limited to exposing nodules to ^{15}N for periods of a few minutes followed by fractionation into several component parts by high-speed centrifugation. In this way, the ^{15}N contents of the symbiont membrane systems and soluble cell material have been calculated (BERGERSEN, 1960). Higher ^{15}N content of the membrane systems than of the soluble cell material can be taken as indirect evidence that fixation is a membrane process. The available data, however, cannot be held to confirm this (see BURRIS, 1966).

Nitrogen fixation by legumes has usually been regarded as a unique symbiotic process—largely on account of the continued failures to demonstrate fixation by the isolated micro-symbiont. It is clear that any site of fixation other than the symbiont membranes would be highly atypical of symbiotic systems. If, however, it is proved to be within the bacteroid cells, then a closer look at the apparent inability of free-living rhizobia to fix nitrogen is indicated.

The site of fixation in non-legume nodules is equally uncertain. By analogy with the legume-*Rhizobium* symbiosis and the lichen symbiosis, it is possible that fixation in non-legumes is also confined to the symbiont membranes, provided that the micro-symbiont does not fix nitrogen in the asymbiotic state. Although there have been reports of nitrogen fixation by free-living actinomycetes and by the micro-symbiont isolated from *Alnus* nodules (see BURRIS, 1966), these have not been confirmed.

Fractionation of non-legume nodules has not yet been achieved, but nodule homogenates have been shown to reduce gaseous nitrogen when supplied with a suitable reducing agent. We can deduce from this that, in the intact nodule, electrons are donated to the process from one or other of the symbionts. The ultimate source must be the macro-symbiont because it is the autotrophic partner of the symbiosis.

The path followed by nitrogen from the gaseous form to the combined form—eventually protein in the symbionts—has been the subject of intense investigation for many years. One cardinal discovery has been the occurrence of a species of haemoglobin (leghaemoglobin or legoglobin) as an apparently unique production of the symbiosis, and its participation in the reduction of nitrogen. It appears in nodules concurrently with the onset of nitrogen fixation, and is thought to be associated with the transference of electrons which produces ammonia as the first stable product of nitrogen reduction.

A view currently held of the symbiotic process (ABEL, 1963) is that gaseous nitrogen becomes bound to haemoglobin functioning as part of a nitrogenase enzyme system. It is then progressively reduced, through the diimide ($HN{=}NH$) and hydrazine ($H_2N{-}NH_2$) stages to yield ammonia. Only at this final reductive stage is the bond between the haemoglobin and nitrogen thought to be broken.

Perhaps the best support for this view is the rather negative, but significant, evidence that no intermediate of prior occurrence to NH_3 has ever been detected in either the symbiotic or the free-living fixation process. But such intermediates may yet be discovered by using the radioactive isotope of nitrogen (^{13}N). There are, however, major difficulties involved in methodology, mainly because ^{13}N has a half-life of slightly less than ten minutes. Counts made at any considerable time after expiry of the half-life naturally have a high attendant error.

From the evidence available at present, we cannot with certainty say that root nodule nitrogen fixation is a unique function of symbiosis. Not that this is of immediate practical significance—but it raises the fundamental point of whether the nitrogen-fixing process can be separated completely from the macro-symbiont and shown to be, enzymatically at least, controlled by the micro-symbiont.

4.3 Adaptive physiological processes

The symbiotic unit, no less than free-living plants, is sensitive to external stimuli, and this is particularly true of symbioses involving unicellular algae. These are examples of coexistence that have resulted in the formation of photosynthetic holobionts and which are therefore dependent for survival upon sensitivity to light. It is a remarkable fact, however, that there is no apparent development of phototropic response in the lichen symbiosis. This can no doubt be explained by the virtual absence of photoresponse by fungal mycelia. But this symbiosis has achieved sensitivity to light, and at the same time has achieved autoregulation of growth and metabolism, by virtue of unique physiological characteristics of the mycobiont cortical cells.

In the other major sphere of algal symbiosis—with aquatic invertebrates—sensitivity to environment has developed along an entirely different line. The zoobiont in its free-living state is a locomotory organism; with the assumption of symbiosis this has been transformed to a locomotory response to light intensity. But the anomalous feature of this symbiosis centres upon the change in mechanism of phycobiont photosensitivity. In the free-living condition there is an inherent locomotory sensitivity to light; this is lost in the symbiotic condition, but is replaced by an acquired photosensitivity of the holobiont.

The physiological mechanisms involved in these processes of metabolic regulation and photosensitivity are closely associated with environmental effects on the cells and with excretions from the cells. We are therefore concerned largely with surface activities of the symbiont cells rather than with endogenous physiological activity. Autoregulation of symbiotic systems can thus be regarded as the fortuitous synchronization of symbiont surface activities resulting in the development of unique systems of control over the gross metabolic activity and behaviour of the holobiont.

4.3.1 Growth regulatory processes

The lichen symbiosis merits special attention with respect to its water relations because the symbionts exist in a delicate state of growth balance which is now known to be largely governed by the amount of moisture, as water or as water vapour, available to the holobiont.

Free-living algae and fungi respond, in their growth, to increasing water status and this applies no less to the two in symbiotic union. The rates of photosynthesis and respiration of many lichens show different responses to change in water status. A lichen thallus exposed to short daily periods of moisture will tend to deplete its store of photosynthetic products by the process of respiration, because this continues at much lower moisture contents than does photosynthesis. That is to say, there exists a minimum level of water content which must be exceeded in order for the symbiosis to show a net gain in dry matter. This is called the moisture compensation point (MCP). Continued exposure to high moisture status, up to the level at which gaseous exchange is limited by the flooding of the intercellular air spaces, results in excessive growth of the phycobiont and consequent breakdown of the symbiosis.

The limitations imposed on the physiological functioning of the symbiosis indicate that the ratio of total time above the MCP to total time below it, is a critical factor in maintaining equilibrated growth. This ratio must obviously be greater than unity.

An important function of water in the physiology of lichens is in the regulation of the amount of light reaching the phycobiont layer. The fungal cortex acts as a light screen or filter. As much as four times the amount of light can pass through the cortical cells of *Peltigera* in the saturated condition compared to the air-dry condition. The screening effect may be demonstrated by moistening the upper surface of any grey thallose lichen with water. The rapid colour change to green is caused by the phycobiont cells showing through the now more transparent fungal cortex.

The mechanism is ingeniously simple. In the saturated condition, the cortical cells are fully expanded or turgid and allow maximum transmission of light per unit surface area. With loss of water, the cells and the intercellular spaces contract; there are now more cells in unit area and less light is transmitted. Not only are there more fungal cortex cells per unit area in the dry state, but there are also more phycobiont cells per unit area because they too contract. This means that a fixed number of phycobiont cells collect less radiant energy than they do when the thallus is saturated, entirely apart from the effect of contraction or expansion of the cortical cells.

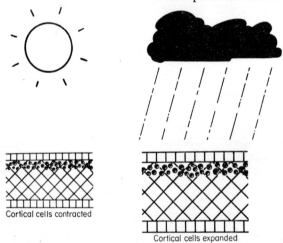

Fig. 4-8 The upper cortex of lichens acts as a light screen. This is a hydro-regulatory mechanism which controls the light transmissibility of the cortical cells.

The moisture effect is thus two-fold. Contraction of the cortical cells with subsequent hindrance of light transmission is supplemented by contraction, on a lesser scale, of the phycobiont cells. This is a self-compensatory system, for the greater the insolation (exposure to sun) the greater the

§ 4.3   ADAPTIVE PHYSIOLOGICAL PROCESSES   47

rate of drying out of the cortex, and thus the lower the amount of light reaching the phycobiont cells (Fig. 4–8). Also, there is a naturally low incidence of high atmospheric moisture coupled with high insolation.

The significance of this control system to the symbiosis lies in the provision of an effective means of equilibrating the metabolic activity of the phycobiont cells to a level that is compatible with maintenance of the symbiotic state. It also prevents the condition of long exposure to high insolation which would be detrimental to the phycobiont cells, for many are known to be intolerant of strong sunlight.

The environmental conditions that ensure optimum symbiotic growth are, therefore, overcast skies with relatively low light intensity but high atmospheric moisture. This is reflected in the geographical distribution of lichens. The centres of highest population density are the middle to high altitude mist belts of the tropics and sub-tropics and similar situations from sea-level upwards in temperate and boreal regions. Water in the lichen symbiosis thus provides for the operation of a hydroregulatory mechanism whereby the symbionts are maintained in growth equilibrium. This is unparalleled by any other known symbiosis.

In many lichens this mechanism of growth control is supplemented by the deposition of coloured lichen acids in the upper cortex (§4.2.1). Familiar examples are the yellow rhizocarpic acid of *Rhizocarpon* and the deep orange physcion of *Xanthoria*. It is commonplace observation that

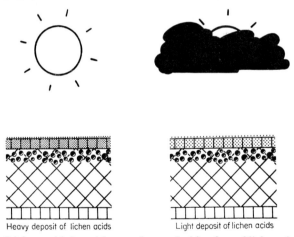

Fig. 4–9   Lichen acid deposition on the cortical hyphae of lichens is a function of light intensity. The deposit acts as a light filter.

lichens, normally deeply pigmented when growing in sunlight, are weakly or not at all pigmented in shade (Fig. 4–9). Species of *Umbilicaria*, for example, are black when exposed to sunlight, but when part of a thallus becomes shaded by other vegetation, new growth is quite green. The effect of light on pigment formation can be readily seen in the brown *Parmelias* such as *P. omphalodes* or *P. fuliginosa*. Overlapping of the squamules gives rise to partial shading; the exposed parts become deep brown while the shaded parts remain grey or green.

Several species of *Peltigera* are usually deep brown when exposed to strong light, but if a branch of a leafy liverwort such as *Lophocolea* grows over the surface of the thallus, the area underneath the branch reverts to the non-pigmented grey condition. Despite the high degree of transparency of the liverwort, the contrast between the shaded grey part of the thallus and the exposed brown part is so well defined as to produce a perfect silhouette in grey, of the liverwort branch (SCOTT, 1964).

The conclusion to be drawn from this frequently observed phenomenon (most species of *Peltigera* show it) is not that the brown pigment is formed in response to light exposure, but that it is bio-destructible. Considering the relative growth rates of the lichen thallus and liverwort branch, it is safe to surmise that the entire thallus was brown before the liverwort grew over it. In partial shade, synthesis of the brown pigment is depressed and we can conclude that the grey silhouette is the result of a more rapid rate of degradation than synthesis of the pigment.

This does not imply that the breakdown products of the pigment are used in any other metabolic sequence. Neither is it evidence that pigment formation and light intensity are correlated through the medium of phycobiont photosynthesis. There could equally well be a light-sensitive reversible system whose equilibrium is displaced towards pigment formation or degradation depending on the intensity or duration of light.

The relationship between light intensity and lichen acid formation is frequently revealed by coloured lichens such as *Xanthoria parietina* growing on tree branches. All gradations can be observed from deep orange thalli to those with scarcely any physcion. Most interesting, however, is the occurrence of thalli growing partly in full sunlight, partly in deep shade, due to the presence of sunflecks and the curvature of the branches. In these thalli there is a clear boundary between orange and green segments. One part of the thallus is able to complete the synthesis of physcion; the other is not and remains green (SCOTT, 1964).

The two systems of growth regulation—fluctuation of moisture content and the deposition of a lichen acid light filter—are partly supplementary in their operation. The additional feature of the lichen acid filter is that it may function in the selective control of light penetration to the phycobiont layer. Coloured lichen acids show absorption peaks in the visible and ultraviolet wavebands. Visible wavelength absorption varies for different acids but is usually towards the blue end of the spectrum. Physcion, for example, shows maximum absorption in methanol solution at 436 mμ (HARPER and LETCHER, 1966). In *Xanthoria* and other species containing physcion, part of the blue region of the spectrum is thus filtered out by the crystal deposit on the cortical cells.

The efficiency of these two growth regulation factors is illustrated by the very sharp zonation of lichens on seashore rocks and notably on the large granite domes of central Africa (SCOTT, 1967). In both habitats, minute differences in substrate elevation, relative to the seasonal mean water level, are sufficient to alter diurnal and seasonal moisture regimes to such an extent that change in species composition can be detected over very short distances.

The preponderance of coloured lichens in these same habitats, where they are regularly exposed to conditions of high insolation and high moisture content from tidal movement or from gravitational water flow, suggests that pigmentation constitutes a selective advantage. There can be little doubt that in these situations the lichen acid filter acts as a supplement to the hydroregulatory mechanism of growth control.

One highly anomalous and as yet unexplained feature of the lichen symbiosis, arising from this consideration of growth control, is the apparently limited translocation of certain metabolites within the holobiont. The clear-cut silhouette of *Lophocolea* on *Peltigera* and the equally clear-cut disjunction line between orange and green parts of *Xanthoria* thalli growing over a shade-light boundary, are quite incompatible with the clearly demonstrated translocation of glucose and ribitol (RICHARDSON *et al.*, 1967) and with the evident exudation of orsellinic acid or its precursors to the mycobiont cell surfaces (MOSBACH and EHRENSVÄRD, 1966). If these events routinely occur, it is pertinent to ask why lichen acid and pigment synthesis ceases so abruptly with change in light intensity over a thallus surface, and why, on a *Xanthoria* thallus in deep shade, the production of physcion continues in the apothecia (§4.2.1). These are but two of the intricate problems of symbiosis that await solution.

4.3.2 *Photosensitivity*

One major failure in the evolution of symbiotic systems towards autotrophism is the apparent lack of development of phototropic sensitivity. The lichen symbiosis has advanced far in its simulation of the angiosperm leaf, yet we can detect little hint of thallus orientation either towards or away from light. But this is an aspect of lichen physiology that has scarcely been investigated.

A very different situation is apparent in the symbiosis between unicellular algae and aquatic invertebrates. Here, we have the unique circumstance of the holobiont acquiring a type of photosensitivity that is characteristic of neither of the participants as free-living organisms. Most phycobionts so far identified as partners of these symbioses do show photosensitivity, but this is lost with the assumption of symbiosis. The sensitivity acquired by the holobiont is the result of nutritional interaction between the symbionts and is expressed as a locomotory response by the zoobiont.

Among the best examples of this acquired sensitivity are the coral polyps, very few of which are asymbiotic, various anemones and bivalves. The virtual confinement of these symbiotic organisms to shallow waters—the euphotic zone—suggests that some mechanism of response to light is involved in their distribution. Such a response has been demonstrated for anemones of the genus *Condylactis* by ZAHL and MCLAUGHLIN (1959). They randomly arranged specimens in a concrete pool, part of which was shaded from full sunlight. Distribution counts after 5 days revealed that two-thirds of the anemones had migrated to the shaded part of the pool. The remainder, in full sunlight, were observed to have their trunks and tentacles more retracted than those in shade. Asymbiotic specimens, obtained by 'bleaching' in darkness for 24 days, showed no response to light intensity in terms of movement or tentacle retraction when subjected to the same treatment.

Clearly this is an example of locomotory response to light intensity induced by the symbiotic condition. What is not so clear is how this response is effected or how, once effected, it is nullified so that the holobiont takes up an optimum position relative to light intensity. Nor is it clear how anemone tentacles, once stimulated to movement, are subsequently maintained in the characteristic position of partial retraction when exposed to full sunlight.

There is no doubt that photosensitivity is a secondary reaction mediated by the interaction of the co-symbionts at the nutritional level. Experimental evidence favours the theory that zoobiont metabolism, particularly in coelenterates which have no excretory organs, is geared to the rate of utilization of waste products by the phycobiont. These include CO_2, phosphates, nitrates, sulphates, ammonia and organic nitrogen compounds. Most of these are stimulatory, if not essential, to phycobiont metabolism; this fact has been postulated by MCLAUGHLIN and ZAHL (1966) to be a factor inducing the free-living phase of the phycobiont to assume symbiotic status. Their assertion is based on the observed low levels of phosphates, nitrates and vitamins in tropical sea water, and on their demonstration that uptake of $^{14}CO_2$ by *Symbiodinium microadriaticum*, the phycobiont of numerous coelenterates, is stimulated when uric acid is supplied in the culture medium.

If this is indeed the mechanism of initiation of the symbiosis (Fig. 4-10), it represents a close parallel to the situation in most other types of symbiosis for which one of the inductive factors is a deficiency of nutrients (§4.1.1). It has the unique feature, however, of the absence of an induction factor governed by deficiency of carbohydrates.

Migration of coelenterates from low to high light intensity is perhaps a response by the zoobiont to an increasing rate of removal of its waste products. When the rate of uptake of these products by the phycobiont, and their diffusion into the growth medium, become sufficiently high to prevent 'constipation', we can envisage that the stimulus to movement *up* the light intensity gradient will cease, and thus that the holobiont will take up a position in the optimum light intensity. Movement *down* the light intensity gradient, as shown by Zahl and McLaughlin's anemones, cannot be explained in these terms. The most logical theory appears to be that zoobiont movement is sensitive in some way to the excretion products of the phycobiont. Since the rate of excretion is directly related to light intensity, it follows that interaction of the responses to the two stimuli, that is to say, removal of zoobiont excretions and accumulation of phycobiont excretions, will result in overall sensitivity to light.

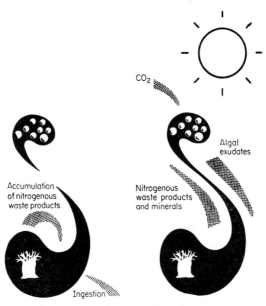

Fig. 4–10 The coelenterate symbiosis is thought to be maintained by the utilization of zoobiont waste products by the phycobiont. In the asymbiotic condition, accumulation of waste products considerably lowers zoobiont vitality.

Photosensitivity in the endozoic symbiosis has the distinction of being a unique attribute, not only because both symbionts acquire it, but also because it is partly controlled by the removal of waste metabolites. It has been acquired by the accident of interaction between excretion products in their effects on metabolic activity of the zoobiont. Still to be explained, however, is how the light stimulus is translated to a movement sequence towards or away from light. Much more experimentation is required, particularly with respect to unilateral light, before we can see the way to an eventual solution of this aspect of symbiosis.

4.4 The integrated symbiotic unit

We can now appreciate the complexity of the metabolic interactions involved in the initiation and maintenance of symbiotic systems. Some of the interactions and processes are unique, such as nitrogen fixation or the acquired photosensitivity of sea anemones; others, such as the interflow of metabolites, are of general occurrence and can be said to be unspecific to any one system, but the prerequisites of all.

The flow of carbohydrate from one symbiont to the other, whether continuously unidirectional, as in the majority of symbioses, or whether bidirectional, as in the phase-separation symbiosis of the green orchids, is the keystone on which the edifice of symbiotic association precariously rests. The supporting pillars are the interchange of nutrients and the regulatory systems serving to maintain equilibrated growth.

No symbiotic system can survive unless the growth and metabolic rates of the partners are accurately tuned in unison. In every symbiosis there are countless failures to strike this unison, but the success of present-day symbiotic association is living proof of the adaptability and of the plasticity of plants in their capacity to form a near-perfect dovetailing of their various metabolic processes.

The symbiotic unit in many respects functions as a single entity. In the more highly developed examples such as in lichens, or *Cyanophora*, the two organisms behave so much as a single physiological unit that it is difficult to escape the conclusion that these are indeed obligate symbioses. But symbiotic organisms fall short of total integration at the point of reproduction. There are familiar examples of the dispersal of algal cells adherent to lichen ascospores, as in *Dermatocarpon*, and of the budding of hydroids with associated phycobiont cells; no system, however, is known in which the sexual phases of reproduction show any degree of integration.

I have branded symbiosis as a physiological makeshift. It is indeed a makeshift, as the many devices that have been evolved in symbiotic systems amply demonstrate. But perhaps most revealing is the often overlooked fact that this makeshift represents a collusion of organisms, at least one of them heterotrophic, which has come part or whole way towards an autotrophic existence. Diverse plants, thrown together in the evolutionary maelstrom, have acquired autotrophism by this makeshift mechanism of symbiosis. Man's testimony to the success of this device is no less than the extent to which he relies on it.

Symbiotic Systems in Nature 5

5.1 Symbiosis in the economy of Nature

Nature, in one sense, is compounded of a series of interdependent cyclic systems—the carbon, nitrogen, phosphorus, and as many other cycles as there are elements utilized by plants and animals. Each of these represents one facet of the interdependence of 'things natural' and, in the context of this chapter, the nitrogen cycle is of special interest.

The impact of symbiosis on the nitrogen cycle only becomes apparent when we see data purporting to represent the annual amount of molecular nitrogen fixed by symbiotic organisms—of the order of one hundred million tons per year. While the true magnitude is beyond assessment, it is nevertheless a sobering thought that, but for the advent of the leguminous plants and the symbiotic association between them and *Rhizobium*, the present-day status of soil nitrogen would indeed be very low—much too low to support anything like the present world cover of natural vegetation, let alone the vast areas under intensive cultivation.

Leguminous crops have for centuries been known to improve soil conditions for other crops, but we still do not make sufficient use of them as a means of converting molecular nitrogen into forms that can be utilized by non-leguminous crops. In terms of efficiency of fixation, the biological process compares favourably with the commercial manufacture of nitrogen fertilizer—a good legume crop should be capable of fixing upwards of 200 pounds of nitrogen per acre in a season's growth. This is roughly equivalent to the application of one ton per acre of NPK fertilizer.

Perhaps less spectacular, but equally important, is the part played by the various mycorrhizal systems that develop in association with conifer and broad-leaved trees. World economy relies heavily on successful afforestation schemes in which conifer species play a major part. We have only to think in terms of the large afforestation schemes that have come into being on so much of the nutrient-deficient marginal land in Britain and other countries, to realize the importance of the mycorrhizal symbiosis in the stimulation of nutrient uptake from poor soils.

The economic aspects of endotrophic mycorrhizal systems remain largely unexplored. Vesicular-arbuscular mycorrhizal fungi are known, in a few instances, to enhance the uptake of phosphate and other nutrients, but we are a long way from being able to assess the practical importance of what has been said to be a virtually universal symbiotic association (MOSSE, 1963). Success has frequently rewarded agriculturists in their efforts to create optimum soil conditions for plant growth, in terms of applied fertilizer and soil physical conditions. What is still the unknown quantity in this work is the participation of the endotrophic symbiosis in the attainment of the high yields now regarded as commonplace.

Although now of minor commercial importance, lichens have in former times assumed a major role in the economy of the world's northern populations. The numerous forms of 'reindeer moss' are still put to a variety of uses ranging from cattle bedding, reindeer and caribou fodder, to the production of antibiotics and for decorative purposes. At least three commercial antibiotic preparations, based on usnic acid, are now available and we can expect more to make their appearance.

The major impact of the lichen symbiosis on world economy is hidden within the leisurely progress of plant succession in terrestrial habitats. We can seldom point to large-scale evidence that lichens are among the world's important plant pioneers—but this they are. Their very resistance to extremes of environment mark them out as 'obvious' forerunners of soil development and mesophytic vegetation cover. They are to be found as the last representatives of plant life in mountains, far above the limits of other vegetation; in the polar regions they are recorded from situations in which it is scarcely conceivable that plant life could exist. In more equable situations, they are no less evident as agents of rock disintegration. This is the great anomaly of the lichen symbiosis—lithophytic (rock inhabiting) lichens destroy their own substrate by alternate expansion and contraction with diurnal and seasonal changes in moisture content. Slow as it is, this continuous process of rock erosion is the contribution of the lichen symbiosis to the economy of Nature.

5.2 The struggle for existence

Symbiosis has often been regarded as a last-ditch stand against the pressures of competition for nutrition and for living space. Non-chlorophyllous symbionts, such as lichen or mycorrhizal fungi, have been described as retrogressive evolutionary lines which have lost the ability to 'fend for themselves' in the heterogeneous milieu of the terrestrial environment. I prefer to think of symbiosis as an evolutionary device that has succeeded admirably in overcoming the disability of heterotrophism.

Evolutionary trends can never be directed away from association. Wherever the situation occurs of a more readily available supply of nutrient material, that is to say, wherever a dominant concentration gradient of nutrient occurs, the physiological organization of the plant decrees that this gradient has a dominant effect on its metabolism. No plant can 'select' its nutrition as can animals, and thus it is inevitable that wherever green and non-green plants by chance grow within mutual influence distance, closer association rather than divergence will be induced by evolutionary loss of function such as of adaptive enzyme systems.

The present-day extent of physiological association between terrestrial plants—including soil micro-organisms—is such that a non-aligned plant is a noteworthy exception. In contrast, and significantly so, planktonic organisms reveal less tendency towards association. Here we have a moving medium of growth which is more highly uniform in every respect that the terrestrial environment. In such a medium, there is less chance of localized nutrient deficits occurring and therefore less evolutionary pressure directed towards an alternative source of nutrition.

It must be recognized that there is no such condition as obligate symbiosis, in the sense that one of the symbionts cannot survive in the free-living state. The distinction between obligate and facultative association, as in the host-parasite relationship, is highly artificial and is dictated only by the lack of evidence for the existence of free-living phases. Taken to the logical conclusion, it must be conceded that, in the light of modern tissue culture developments, all plant organisms as well as all nucleated plant cells are capable of independent existence. The only requirement is technique. But when we confine our consideration of symbiosis to the natural environment, there is no acceptable *rationale* of this phenomenon other than one which recognizes that the participant of lower rank in the scale of autonomy is induced, by force of nutritional circumstances, to dependence upon an associate of higher rank.

In terms of competitive ability, the symbiotic unit, derived from chlorophyllous and non-chlorophyllous organisms, stands out clearly as an advantageous association. The numerous unique physiological characteristics of the lichen symbiosis are fundamental to the extremely wide and varied distribution of lichens. Few other organisms, not possessed of these characteristics, could survive for long periods on rocks whose temperatures range from sub-zero in the polar regions to around $50°C$ in the tropics; nor could they survive at air dryness for more than six months of every year on rocks exposed to full tropical sunlight.

One of the most elegant of symbiotic systems, but yet the least understood, is the plant-animal symbiosis that has developed in more than a hundred genera of protozoa and coelenterates. For some, the ultimate in symbiosis has been achieved—autotrophism; for the majority—photosensitivity. Both achievements clearly vindicate the assertion that biological advantage accrues from the sharing of life with a companion organism.

Future evolutionary trends must always be towards more highly specialized degrees of association in already existent symbioses, and must be towards the initiation of new symbiotic systems. Any heterotrophic organism that produces a mutant capable of extracting nutrition from a source not available to its competitors, is conferred with an evolutionary advantage and has thereby a greater chance of survival than its competitors. Additionally, any mutant of the heterotrophic organism that shows physiological compatibility with a potential co-symbiont is at a further advantage. This is essentially how symbiosis has evolved and it is how symbiosis will continue to evolve.

There are already a few thousand fungi associated with chlorophyllous plants; some destructively parasitic, but many that can be classified as benignly parasitic—extracting their total nutritional requirements from the host but having little effect otherwise. This is the very threshold of symbiosis.

The rhizosphere and phyllosphere floras are further pertinent examples of the possibilities of developing symbiosis. Many of the components of these floras are at a nutritional advantage; it is but a short step in the evolutionary time scale towards more interdependent association, following the paths undoubtedly taken by present-day symbiotic systems.

But there is a great deal of the 'hen or the egg' philosophy surrounding symbiosis; so much so that we must in the end be content with merely stating that evolution has thrown up numerous lines of organisms which have become so specialized in their nutrition that they are at a gross disadvantage in the free-living state, and that specialization is more likely than not to continue.

Thus, in the struggle for existence in the finite volume of space on the earth's surface and in its oceans, the assumption of symbiotic status has provided the means whereby a variety of organisms has acquired a new lease of life. For them, the sands of time are still running: from them we can perhaps attain a greater degree of sensibility towards the life of which we ourselves are part. With time, we shall surely grasp the true implications of organismal association, in the knowledge that no form of life can exist completely divorced from association with some other form. In this last analysis, symbiosis becomes synonymous with life.

References

ABEL, K. (1963). *Phytochem.*, **2**, 429–435.
AHMADJIAN, V. (1966). *Science, N.Y.*, **151**, 199–201.
AHMADJIAN, V. (1967). *Phycologia*, **6**, 127–160.
BERGERSEN, F. J. (1960). *J. gen. Microbiol.*, **22**, 671–677.
BERGERSEN, F. J. and BRIGGS, M. J. (1958). *J. gen. Microbiol.*, **19**, 482–490.
BJÖRKMAN, E. (1949). *Svensk. bot. Tidskr.*, **43**, 223–262.
BJÖRKMAN, E. (1960). *Physiologia Pl.*, **13**, 308–327.
BOND, G. (1959). *Advmt. Sci., Lond.*, **15**, 382–386.
BOND, G. (1963). The Root Nodules of Non-leguminous Angiosperms. In *Symbiotic Associations*. Edited by P. S. NUTMAN and B. MOSSE. 13th Symp. Soc. gen. Microbiol. Cambridge University Press, London.
BOND, G. and SCOTT, G. D. (1955). *Ann. Bot., Lond.*, N.S. **19**, 67–77.
BURRIS, R. H. (1966). *A. Rev. Pl. Physiol.*, **17**, 155–184.
DAFT, M. J. and NICOLSON, T. H. (1966). *New Phytol.*, **65**, 343–350.
DAVIS, B. D. (1950). *Experientia*, **6**, 41–50.
DE BARY, A. (1879). *Vers. Deut. Naturforscher und Ärzte zu Cassel*. Tagebl. 51, Strassburg.
DOWNIE, D. G. (1959). *Trans. Proc. bot. Soc. Edinb.*, **38**, 16–29.
FOSTER, R. C. and MARKS, G. C. (1967). *Aust. J. biol. Sci.*, **20**, 915–926.
FURMAN, T. E. (1966). Abstract in *Am. J. Bot.*, **53**, 627.
GOREAU, T. F. and GOREAU, N. I. (1960). *Science, N.Y.*, **131**, 668–669.
HALE, M. E. (1967). *The Biology of Lichens*. Edward Arnold, London.
HALL, W. T. and CLAUS, G. (1963). *J. Cell Biol.*, **19**, 551–563.
HARLEY, J. L. (1959). *The Biology of Mycorrhiza*. Leonard Hill, London.
HARPER, S. H. and LETCHER, R. M. (1966). *Proc. Trans. Rhod. scient. Ass.*, **51**, 3–31.
HESS, D. (1959). *Z. Naturf.*, **14b**, 345–347.
HILL, D. J. and WOOLHOUSE, H. W. (1966). *Lichenologist*, **3**, 207–214.
JACKSON, R. M. and RAW, F. (1966). *Life in the soil*. Studies in Biology No. 2. Edward Arnold, London.
KINUGAWA, K. (1965). *Bot. Mag., Tokyo*, **78**, 366–373.
LAMB, I. M. (1951). *Can. J. Bot.*, **29**, 522–584.
LAWRENCE, D. B., SCHOENIKE, R. E., QUISPEL, A. and BOND, G. (1967). *J. Ecol.*, **55**, 793–813.
LOCHHEAD, A. G. and BURTON, M. O. (1957). *Can. J. Microbiol.*, **3**, 35–42.
MCLAUGHLIN, J. J. A. and ZAHL, P. A. (1966). Endozoic Algae. In *Symbiosis*, Vol. 1. Edited by S. M. HENRY. Academic Press, New York.
MELIN, E. and NILSSON, H. (1957). *Svensk. bot. Tidskr.*, **51**, 166–186.
MEYER, F. H. (1966). Mycorrhiza and other Plant Symbioses. In *Symbiosis*, Vol. 1. Edited by S. M. HENRY. Academic Press, New York.
MORRISON, T. M. and ENGLISH, D. A. (1967). *New Phytol.*, **66**, 245–250.
MOSBACH, K. (1964). *Acta chem. scand.*, **18**, 329–334.
MOSBACH, K. and EHRENSVÄRD, U. (1966). *Biochem. biophys. Res. Commun.*, **22**, 145–150.

MOSSE, B. (1963). Vesicular-arbuscular Mycorrhiza: an extreme form of Fungal Adaptation. In *Symbiotic Associations*. Edited by P. S. NUTMAN and B. MOSSE. 13th Symp. Soc. gen. Microbiol. Cambridge University Press, London.

MUSCATINE, L. and HAND, C. (1958). *Proc. natn. Acad. Sci. U.S.A.*, **44**, 1259–1263.

MUSCATINE, L. and LENHOFF, H. M. (1963). *Science, N.Y.*, **142**, 956.

NUTMAN, P. S. (1952). *Ann. Bot., Lond.*, N.S. **16**, 79–101.

RICHARDSON, D. H. S., SMITH, D. C. and LEWIS, D. H. (1967). *Nature, Lond.*, **214**, 879–882.

SCHWENDENER, S. (1869). Die Algentypen der Flechtengonidien. *Programm für die Rektoratsfeier der Universität Basel.*

SCOTT, G. D. (1960). *New Phytol.*, **59**, 374–381.

SCOTT, G. D. (1964). *Advmt. Sci., Lond.*, **20**, 244–248.

SCOTT, G. D. (1967). *Lichenologist*, **3**, 368–385.

SCOTT, N. S. and SMILLIE, R. M. (1967). *Biochem. biophys. Res. Commun.*, **28**, 598–603.

SILVER, W. S., CENTIFANTO, Y. M. and NICHOLAS, D. J. D. (1963). *Nature, Lond.*, **199**, 396–397.

SLANKIS, V. (1958). The Role of Auxin and other Exudates in Mycorrhizal Symbiosis of Forest Trees. In *The Physiology of Forest Trees*. Edited by K. V. THIMANN. Ronald Press Company, New York.

SMITH, S. E. (1966). *New Phytol.*, **65**, 488–499.

STEWART, W. D. P. (1966). *Nitrogen fixation in Plants*. Athlone Press, London.

STEWART, W. D. P., FITZGERALD, G. P. and BURRIS, R. H. (1967). *Proc. natn. Acad. Sci. U.S.A.*, **58**, 2071–2078.

THIMANN, K. V. (1963). *The Life of Bacteria*, 2nd edn. The Macmillan Company, New York.

TRAPPE, J. M. (1962). *Bot. Rev.*, **28**, 538–606.

ULRICH, J. M. (1960). *Physiologia Pl.*, **13**, 429–443.

WALLROTH, F. W. (1825–1827). *Naturgeschichte der Flechten*, Vols. 1 and 2. Frankfurt-am-Main. Cited by SMITH, A. L. (1921). *Lichens*. Cambridge University Press, London.

WEBSTER, S. R., YOUNGBERG, C. T. and WOLLUM, A. G. (1967). *Nature, Lond.*, **216**, 392–393.

ZAHL, P. A. and MCLAUGHLIN, J. J. A. (1959). *J. Protozool.*, **6**, 344–352.

MUNITY COLLEGE LIBRARY

AS QK 918 .S33 1969 c.1

Scott, George D.
   Plant symbiosis

DISCARD

LANSING COMMUNITY COLLEGE LIBRARY
LANSING, MICHIGAN